EDWARD
The First African-American Doctorate
BOUCHET

T0204249

EDWARD
The First African-American Doctorate
BOUCHET

Edited by

Ronald E. Mickens
Clark Atlanta University, USA

World Scientific
New Jersey • London • Singapore • Hong Kong

Published by

World Scientific Publishing Co. Pte. Ltd.

P O Box 128, Farrer Road, Singapore 912805

USA office: Suite 1B, 1060 Main Street, River Edge, NJ 07661

UK office: 57 Shelton Street, Covent Garden, London WC2H 9HE

British Library Cataloguing-in-Publication Data
A catalogue record for this book is available from the British Library.

EDWARD BOUCHET – THE FIRST AFRICAN-AMERICAN DOCTORATE

ISBN 981-02-4909-8 (pbk)

Printed by FuIsland Offset Printing (S) Pte Ltd, Singapore

This volume is dedicated to the memory of my friends, colleagues
and fellow scientists ...
Howard Foster
Warren E. Henry
James R. Lawson
Henry C. McBay.

Preface

The first African American to receive the doctorate in any field of study was Edward A. Bouchet. This degree was granted to him by Yale University in 1876 in the area of experimental physics. The primary purpose of this small edited volume is to provide some insight on his life, career, and the social environment in which he lived and worked. However, a difficulty in trying to fully understand and analyze the actions of Bouchet at various stages of his life lies in the almost complete lack of personal records. There is no evidence that any collection of his writings exist, except for a few short letters and the statements he prepared over the years for the Yale University Biographical Record of the Class of 1874. This means that contradictory statements and varying interpretations of events occurring in his life arise. Consequently, a second goal of this volume is to have one or more readers become so excited by the prospect of wanting to understand Bouchet that they will then carry out the thorough, deep study required for this effort.

This book consists of five chapters and a number of appendices with materials related to Bouchet's influence on later generations of African Americans. The first chapter provides a brief history of African American life in New Haven, Connecticut, and the impact of Yale University on this community in the decades leading to Bouchet's arrival at The College. How Bouchet got to the Hopkins Grammar School and the intimate details on how he was received remain largely a mystery. The second chapter deals with these issues based on the limited information currently available. The bulk of Bouchet's life and career was spent at the Institute for Colored Youth (ICY) in Philadelphia. Chapter 3 discusses Bouchet within the context of his work at the ICY, his impact on the African American community in Philadelphia, and follows his various "careers" after the closing of the ICY. In chapter 4, the writer poses, but cannot answer, a number of questions related to Bouchet in his capacity as a highly trained individual who was forced by historical circumstances to never receive a position of responsibility commensurate with his high academic achievements. Finally, chapter 5 reviews the influence that African Americans have had on the scientific community in the United States with a particular emphasis on the limitations resulting from the lack of equitable educational opportunities and funding for research.

The materials in the six appendices cover a variety of subjects. Appendix A reproduces a copy of a letter to me from a former middle school pupil of Bouchet. It took nearly a century before the first African American woman achieved the doctorate in physics. This person, Willie Hobbs Moore, received her degree in 1972 from the University of Michigan. A brief summary of her life and career is given in Appendix B. The second African American to receive a Ph.D. in physics was Elmer Samuel Imes (1918, University of Michigan). He was one of the first African American scientists to attain international fame for research. His work centered on the precision measurement of the rotation-vibration spectra of certain diatomic molecules.

In the 1970's, black scientists began to organize themselves into professional societies outside of the existing mainstream national discipline based scientific organizations. The fundamental reason for this course of action was the need for organizations "in which Blacks play a major role in creating and developing activities and programs themselves for themselves." These new entities were not in conflict with either the goals or the missions of the mainstream professional organizations and were not intended to supplant any of them. Appendix D presents a short history of the National Society of Black Physicists.

In June 1988, a meeting was held in Trieste, Italy at the International Centre for Theoretical Physics (ICTP). This event was called the First Edward Bouchet International Conference on Physics and Technology. It was a joint effort of ICTP and Black American Friends of ICTP (BAF/ICTP), a support group created to organize this gathering. The central motivation for the meeting was to provide Black physicists from America and African physicists "an arena in which they can share their research results; discuss current topical issues in physics; address problems of mutual concern; and create a continuing organization." The major result coming out of this meeting was the formation of the Edward Bouchet-ICTP Institute. (The current name is Edward Bouchet-Abdus Salam Institute.) The early history of the Institute and its Constitution are the subject of Appendix E.

Appendix F provides a selected list of books, articles, and related publications on African American scientists.

One measure of the significance of an individual is whether major awards, fellowships, etc., are named for them. There are many such honors for Bouchet. They include the recognitions noted below:

* The National Conference of Black Physics Students bestows an award at its annual meeting, the Edward A. Bouchet Award, for "outstanding contributions to teaching, mentoring of students, and service to the minority science community."

* The American Physical Society (APS) presents the Edward A. Bouchet Award annually to "distinguished minority physicist who has made significant contributions to physics research .... The award consists of a stipend of $3,500 plus support for travel to an APS meeting where the recipient will receive the award and give a presentation." The award was established in 1994 by the APS Committee on Minorities in Physics and is supported by a grant from the Research Corporation.

* The Edward A. Bouchet Outstanding Achievements Award is given annually by the Sigma Pi Phi-Beta Tau Boulé "to a rigorously selected group of high school students in the New Haven area who are preparing to graduate. These individuals are nominated for this award by their school principals, headmasters, guidance counselors and members of Beta Tau." The award was first given in 1986 and

includes a check for $500.00, a certificate, and a medal.

* The Edward A. Bouchet Undergraduate Fellowship Program is given at the Yale College and is underwritten by the Office of the President. This competitive fellowship is intended to encourage minorities to pursue doctoral degrees which lead to careers in academia.

Finally, I wish to thank The Alfred P. Sloan Foundation for financial support of this publishing project. I also thank the Fisk University Library Special Collections, the Gale Group, the family of Willie Hobbs Moore, and the Yale University Library for granting permission to publish photographs and other documents included in this volume. As always, I am particularly grateful to Annette Rohrs for electronically processing the various manuscripts and other materials from which this book was formed. Both she and my wife, Maria, provided excellent editorial assistance and encouragement.

Ronald E. Mickens
Atlanta, Georgia
December 2001

Contents

EDWARD

The First African-American Doctorate

BOUCHET

Chapter 1

Early African American Presence in New Haven and Yale University

Curtis L. Patton*

In remarkable ways, Edward A. Bouchet began to be an important part of Yale long before his birth; he continues to be a part of this University and the lives of students and faculty here and abroad who know about him and accept their own chance and occasion for serious study. No words of mine can add to the brave record of his life, his fine scholarship and challenges during a time accurately described as bad for black folk. I am ever grateful to Professor Lee Lorch for sharing the story of Bouchet with me when I was a freshman at Fisk University. Having accepted in full the offerings of a fraternity rush the night before, I had fallen asleep in math at 9:00 AM the following morning. In response to Professor Lorch's question and still under the influence, I failed to respond appropriately. Dr. Lorch stopped his lecture. Staring at me, I thought, he began: 'A century ago today, a boy named Edwards Alexander Bouchet was born in New Haven, Connecticut. He rose above all obstacles to become the first African-American graduate of Yale College, the first admitted to Phi Beta Kappa, the first to earn the Ph.D. in the Americas and was the sixth person of any ethnic group to earn the Ph.D. in Physics in the western hemisphere. Thanks Lee. It does in no way compensate for my performance in your memorable course, but this one is for Bouchet and you.

*Yale University Medical School, Department of Epidemiology and Public Health, New Haven, CT 06520 (`curtis.patton@yale.edu`).

The Educated American Black

Distinguishing African American black people as good or bad Negroes seemed sufficient during the 19th century. If white people made the effort to go beyond such appearances, in the interest of discovering if social gradations among American black folk existed, they might have discovered to their surprise that there was a rather well defined social hierarchy. By inspection, the mass of blacks belonged at the bottom of the American social pyramid. Nevertheless, a small, but growing, upwardly mobile class struggled out of the depths of poverty and/or bondage during that period and struggles still. To liberal whites, this ambitious class appeared then and now to be rewarded for hard work, thrift, and righteous conduct, rewards achievable even among a proscribed people. Based primarily on their first-hand observations, rather than close objective analyses, liberal whites, concluded that there was a black elite, small in number, and often light in complexion whose culture and style of living more closely resembled that of the "better class of whites" than that of the masses of their darker fellows. For those who bothered to investigate, the small group of blacks at the top of the pyramid, "the colored aristocracy," aroused then and now, curiosity and posed challenges. While the distance between the colored aristocracy and the white upper class was as for sure greater than the gulf that separated the former from most blacks, the discovery of such an elite group prompted some observers, in the second half of nineteenth century at least, to question the common view that all blacks were much the same. The acquisition of what might be called culture sharpened distinctions between free people of color and slaves in New England and elsewhere. Assimilation, good behavior and education were included in the recipe for eliminating racial prejudice directed at freedmen, or so they thought. In New Haven, blacks at the top of the pyramid displayed values and life styles of New England ideals of home, church, learning and society. Those who were a part of this colored aristocracy in post-Civil War America inherited from their antebellum antecedents ideals, traditions and patterns of behavior which they hoped set them apart from other blacks. Central to this legacy was an emphasis on behavior and education.

While even the most exceptional black people were considered inferior to whites in the 19th century, this "black upper class" justified its claims to a favored status on various grounds, including record of achievements, status as antebellum free people of color however recently acquired, culture meaning assimilation, education, and, to a lesser degree, wealth. They viewed themselves almost in Darwinian terms as the product of natural selection from which they and their issue emerged as the strongest and fittest of the race, standing as it were in sharp contrast to those who belonged to the greater body. The coming out of this class was ongoing, though small in size. It prompted mixed impulses of self-respect and resentment in the black community. Before W. E. B. Du Bois expressed the notion that black people, like other ethnic groups, ought to benefit from the achievements of their exceptional fellows, the colored aristocrats had subscribed to and practiced the idea with blessed assurance.

The New Haven-Yale Complex: Black People, Church and Events

While there are good reasons to see New Haven as a city with is own character, it is in fact a University City complex, a mingling of spaces, people, ideas, enterprises and events. But it is essential to re-imagine Yale and New Haven as indivisible in order to place Bouchet in context. He belongs at Yale and in a sense is at Yale before his birth, and he continues at Yale after death.

The most casual examination reveals glory, honor and imperfections in the City, the University and their union, a union bound together by faculty, students who come and go and permanent citizens, a place where mixed impulses concerning democratic values and racial equality flourished from the very beginning and flourish still, an environment becoming a place more splendid in which to live and learn, still. Only after being screened through the lens of the single eye of City and College will Bouchet be seen clearly.

Yale, a national resource, an institution devoted to scholarship, has a history that includes awesome and ample examples of involvement in American affairs of state, international affairs and the affairs of black Americans. Yale, an effective if flawed commander in the education of major American figures, dare not shrink from its history, especially in this tercentennial year. Some of who learned and taught in this place dared to explore and express revolutionary ideas about society, religion and science, and some dared with significance to be committed to democracy, freedom and uplift of all Americans and some did not.

During times colonial, Africans arrived as slaves in New Haven and elsewhere in the Americas. Gradually freed by the provisions of the Connecticut's Gradual Emancipation Act of 1784, the black population in the city of Elms grew through reproduction and migration. Almost a century before the great migration of blacks from South to North, William Frances Bouchet joined a planters son, John B. Robertson of Charleston, South Carolina as his body servant at Yale. When Robertson began his studies at The College in 1824, the population in New Haven included little more than 600 blacks and few among them were slaves. Blacks lived separately from whites; the latter, by established tradition, did not approve of independent, well dressed, self-respecting educated blacks; and did not permit social mingling between the races.

By 1850 there were almost 1,000 New Haven blacks whose lives were full of difficulties for the most part. Although the first half of that century bore witness to economic benefits in general, well paying jobs were very few for African Americans. About half worked as servants to the wealthy white aristocracy. The other half worked in all sorts and conditions of jobs. A few had the good fortune and opportunity to develop skills and work as artisans in shoemaking, woodworking and other trades while others worked aboard ocean going vessels. A small number worked in the service sector as barbers and waiters, but most blacks were day laborers randomly assigned menial tasks for little reward. Members of the very best African American families in New Haven held the service jobs.

Imagine the familiar: "they find it difficult to obtain houses, of any description." After all the white people are accommodated, down to the poorest and lowest, blacks took the remainder segregated into three of four isolated and undesirable parts of the city. Most lived in "New Liberia," a run down area on lower Chapel Street between the old waterfront and the factories. The living conditions in New Liberia were severe.

One family, the first in which inquiry was mad, occupied an under-ground [sic] room of a very old building. It has a huge fireplace, a coarse floor, and was poorly lighted; one of its wall was a rude stonework built against the bank, upon which the woman said, drops of water stood in damp weather—and sometimes, after heavy rains, the water soaked through and partially inundated the floor; the other sides of the room were of plank and unplaistered [sic]—there was no ceiling over head, but a floor so open that it was easy to converse with the family above, and whenever the upper family had occasion to use the broom, the dust fell through so freely that the same operation became necessary below, —their clothes, bed, and especially heir dishes, when the table was set for meals, each receiving a portion of the descending particles.

New Liberia was known as a center of vice to the God-fearing people of the city; whiskey flowed cheaply and fights were common. The better off black families fled New Liberia moving to other parts of the city, a few to Negro Lane, a part of State Street east of the original Nine Squares. Many moved to Congress and Washington Streets, known then and now as "The Hill." Later on, some settled in an area around "Poverty Square" recognized today as the area around Dixwell Avenue.

There were among the white majority those whose thoughts, words and even deeds supported the peculiar institution of American slavery and the racist creed that authorized it. Members of the lowest caste, New Haven blacks were not passive. There were among them men and women of pride and purpose, dedicated to improving their lot and contributing to the larger society. With the help of people of good will, including a few courageous whites, they constructed as progressive a society as the majority allowed.

From James Hillhouse, 1773, leader of the anti-slavery movement in the First Federal Congress, to Josiah Willard Gibbs, 1809, who befriended the captives of the Amistad, there was concern among the Yale faculty and student body about the lot of African American slaves in general and New Haven blacks in particular. Many in New Haven, including such notables as Theodore Dwight, Jonathan Edwards and Noah Webster, shared New England's fundamental protestant Puritan background, and as such they viewed slavery as a sin. Out of this abolitionist framework there developed an honest interest in the improvement of life for blacks in New Haven. Much of this interest centered on one man, Simeon Smith Jocelyn, artisan by trade and a member of the Center Congregationalist Church on the Green in New Haven. He was a young idealist and around 1820, affected by the plight of black folks in New Haven, he sought to improve their lot through religious education. A few blacks were members of the city's churches, where they were "tolerated rather than welcomed." The rest of the blacks, "... were absolutely without moral and religious instruction and nobody seemed to care." Barred from voting, barred from schools and not really welcomed in local churches, blacks looked among themselves for resources. A few whites of good will encouraged black solidarity. In Jocelyn's words:

> The colored population was then nearly seven hundred souls; many of them living in neglect of public worship, and a large number were ignorant and vicious. There were, however, a goodly number who longed fort the moral and religious improvement of their brethren.

To bring about this religious improvement, Jocelyn began by holding private religious services in his home with the "prominent African-Americans." This group, composed of four black men and seventeen black women, became known as the African Ecclesiastical Society. The first clerk of the new group was man named Prince Duplex, a well-known African American and Revolutionary War veteran who lived in a house "Fronting Easterly on the New Green," probably the current Wooster Square. Jocelyn and the group purchased a small run down building on Temple Street between Crown and George to serve as a church and he served without ordination as the minister.

In 1829, the Temple Street Church made real the black community's hopes for their own institutions. However, the birth of this church suggests a continuing thread between autonomous black groups and a dependency on white people, their standards and support. The initial members had to pass a test administered by Center Church before gaining recognition by the white entity as a separate religious organization. The other New Haven churches agreed to recognize only the most elite and agreeable blacks in their congregation as part of the new Temple Street Church. The new church goals appeared to have been consistent with the Reverend Thomas Brainerd's ambitions for a church in society to bring "the practical learning of schools to the people" by providing an educated clergy to "stimulate friendships among the best of society," in the present case the best of the black society, to raise the "rude and degraded" to "neatness and good order" by well-regulated meetings; to advance industry, economy, moderation and temperance; and finally, "to enforce all the laws and customs by the moral surveillance of the eye of God."

The group grew quickly in its first year; starting with about twenty-five, more than a hundred people soon crowded into the little space on Sunday. On August 25, 1829 the congregation was officially recognized by the Western Association of New Haven County and became the Temple Street Congregational Church, the first African-American Congregational Church in the nation. The Temple Street church moved in the 1960s and become the Dixwell Avenue Congregational Church where it continues with vigor to this day.

Although they suffered much abuse at the hand of some white New Haveners, the African Ecclesiastical Society did have friends in powerful places, among them The Reverend Leonard Bacon, Yale College Class of 1829, pastor of the influential Center Congregational Church on the Green. He was Jocelyn's pastor and the conscience of New Haven. A member of the Underground Railroad, Bacon held strong antislavery views and great interest in the plight of African Americans. He gave an impassioned sermon on July 4, 1826 entitled "Plea for Africa." He held a meeting later in the month to find out what could be done to help Jocelyn. Among those he called upon was his friend and classmate, Theodore Dwight Woolsey, a tutor at Yale College and an abolitionist. Woolsey was a member of a family with strong Antislavery sentiment, a relative of Jonathon Edwards and Theodore Dwight, both of whom spoke out against slavery. Most important to Yale's legacy to this day was Woolsey's interest in the education of African Americans, an interest he held throughout his life. The meeting would have a powerful impact on the life of African Americans in New Haven. The Reverend Bacon, Woolsey and other men of good will founded two organizations, the Antislavery Association and the African Improvement Society. The goals of the African

Improvement Society, as stated in its constitution, were to, "... improve the intellectual, moral and religious condition of the African population of this city, and especially of the United African Congregation," a.k.a. The African Ecclesiastical Society. The African Improvement Society focused their energies on Jocelyn and his congregation. With backing from Bacon Woolsey and the Society, the Temple street Church made remarkable progress. Sunday school and bible classes were begun along with a Temperance Society, a major force in The City for the next several decades. They founded a small day school for black children, the first in the city. It was supported by public funds for six months of the year and for the remainder by private donations, Vashti Duplex, daughter of Prince Duplex, the first clerk of the church, served as the schoolteacher, the first black schoolteacher in New Haven. This would become Edward A. Bouchet's first school.

Times would soon change. In May of 1831, Simon Jocelyn developed the idea of starting an African American College in New Haven. The goal was to educate African Americans in agricultural and technical skills. Jocelyn received wide support from New Haven and national leaders. In August of 1831, Nat Turner led a slave rebellion in Virginia with repercussions across the nation. The general public, already wary of the betterment of African Americans before the Nat Turner rebellion, was violently opposed to any and all agenda in support of blacks. Plans for this college, which would have been the first dedicated to the education of blacks, were squelched by people at Yale, city leaders and by a hostile press. Mob violence racked the city. Blacks were attacked in the streets. Jocelyn was forced to leave the pulpit at the Temple Street Church in 1834 and in 1837, he fled to New York to escape personal harm.

This anti-African American Sentiment continued in New Haven until in 1839 the schooner Amistad was discovered in the Long Island Sound. The ship had sailed from Cuba with a load of 54 Africans. The Africans, lead by a man named Cinque, sized control of the ship and forced the crew to commit to their return to Africa. The crew, realizing that their captors only knew the direction of their home from the position of the sun, tricked the Africans by steering the ship towards Africa by day and pointing it back to America at night and under overcast skies. In this fashion the crew managed a zigzag course into United States waters, the Long Island Sound, where they were spotted and boarded by authorities.

Since the Africans did not speak English, at first only the Amistad crews side of the story was understood. The Africans were charged with piracy and ordered to stand trial. Thanks to an intrepid Yale Professor, Josiah Willard Gibbs, an interpreter was found. The details of the mutiny, as told by the defendants, was revealed, and spurred by New England abolitionists, an outpouring of public support erupted. People of New Haven and all over New England rose to their defense. Queen Victoria of England also pleaded for the Africans. Roger Baldwin, of New Haven represented the Africans. The case went to the U.S. Supreme Court, where former President John Quincy Adams, an abolitionist, defended them. The court ruled in favor of the captives and those who still lived were freed to return to their homeland.

Pioneering Americans at Yale

Due to the Amistad incident, the people of New Haven were now more amenable to the idea of social improvements for black Americans. Baldwin was elected Governor of Connecticut by the anti-slavery vote in 1844. In 1846 the Connecticut Assembly removed references to race in the states constitution. The post-Amistad period was for African Americans a progressive time. They built new houses, bought property, saved money and help provide for their poor and their young people. The African American church in New Haven grew in number and substance and a fledgling upper class developed among blacks in the city. New Haven and Yale enjoyed a resurgence of African American sympathy. On August 19, 1846, Theodore Dwight Woolsey, a founder with Leonard Bacon of the New Haven African Improvement Society, was chosen by the Yale Corporation to become the ninth President of Yale College. Alexander Crummell, the great Pan-Africanist attended Yale for a short time in the 1840s, and Ebenezer Bassett, the first black diplomat, took class at The College from 1854-1855, but neither was officially enrolled; neither had the privileges and responsibilities of Yale students; and neither left Yale with a degree.

Yale was the largest college in America during this time. Long recognized as a seat of power and influence, Yale educated sons of the most prominent families from the north and south. Events at Yale were eagerly watched and discussed throughout the nation, then as now. With Woolsey as president, Yale became a center for abolitionist thought and debate. James Hamilton, a Southern student, published an article in the *New Haven Register* in which he discussed the prevailing sentiment at the college:

> But, within the last few months, Yale has caught the infection, and now raises her official hue and cry against Slavery, as an "unjust institution," and does reverence to the supremacy of the "higher law" ... not, indeed, through public channels, but through the professional chair, she seeks to instill into the mind of the youth entrusted to her care, a detestation for the institution of Slavery, a contempt for those who sustain it and a hostility to the Constitution which sanctions it.

Hamilton writes:

> ... in a series of lectures, ..., by the President ... he has taken great pains to dwell upon the "injustice of Slavery," and our obligations to a "higher law." In order to bring the subject before the great body of students, he has within the last week, ... given us a question for a prize debate before one of the Societies: "Ought the Fugitive Slave Law to be Obeyed?" He has taken occasion to congratulate himself upon the inefficiency of this law ...
> and he continues: Men retire from the lecture room ... some indignant and enraged ... some, with painful surprise ask: – "What can the President mean by the course he is pursuing?" – others, elated with the sanction of such high authority, unscrupulously re-echo the doctrines there promulgated.

The article in the *New Haven Register* by Hamilton was reprinted in the South and caused controversy among Yales Southern alumni. Abolitionist sentiment increased at The College and its institutions. The Divinity School had permitted James W.C. Pennington, a black student from New York, to attend classes as far back as 1834. He was not allowed to participate in any way at the school because he Connecticut Assembly prevented by law out of state African Americans from attending any Connecticut school without permission from the local town. Pennington's presence at Yale followed the failed attempt to establish in New Haven the first black college. Permission for Pennington to enroll at Yale was not expected and not given by the city of New Haven. To skirt the state law, he Divinity School quietly allowed Pennington to audit classes without library privileges. He prevailed and earned international praise as a scholar and Divine, receiving an honorary doctorate from the University of Heidelberg.

The Medical Institution at Yale, founded in 1810, may have had among its faculty and students abolitionists or colonization sympathies. Dr. Ezekiel Skinner, a physician from Ashford, Connecticut, set sale for Liberia on June 21, 1834 under the auspices of the American Colonization Society. During his time in Liberia, he served as Governor of the Colony. Upon his return to America in 1849, Yale bestowed upon him an honorary M.D. for his contributions to the founding of Liberia. Although Dr. Skinner was thought by one author to be of African descent, there are no documents to support his claim. Since his service as a private and a surgeon in the American military during the War of 1812 and his appointment as Governor of Liberia, it is reasonable to assume he was not an African American. That Yale awarded him a degree suggests Yale faculty and administrators favored freedom for African Americans. It is significant to this writer that Charles Hooker, then Dean of Medicine, was a member of the Yale College Class of 1820 and a good friend of Theodore Dwight Woolsey, President of Yale and The Reverend Leonard Bacon, Yale Cooperation Fellow. While Dr. Hooker was a life resident of Connecticut and friends with these powerful men at Yale whose views against slavery were well known, his views on slavery and abolition are not documented. That he found the company of Bacon and Woolsey agreeable may suggest that he shared their views and activities.

During the first half of the nineteenth century, during the time when Bouchet's father was the body servant of Robertson at Yale, great changes were taking place in New Havens social and political structure. Robertson and William Bouchet made their homes in New Haven. They each married and established families in the city. By the time Edward A. Bouchet was born in 1852, change was in the air. African Americans, with help from a small core of determined and influential men, had hope. In spite of periods of opposition, strong and violent, sentiments were changing. By the mid 1850's, moderate abolitionists controlled the state government. Yale College and President Woolsey were in the forefront of movements in support of blacks and the professional divisions of The College followed suit. Important elements had come together to produce an interesting climate, a climate in 1854 during which the first African American, Cortlandt Van Rensselaer Creed, M.D. 1857, was admitted to study medicine at Yale. Born and raised in New Haven, a life long resident, Cortlandt Creed was not subject to the "Black Law" that prevented Pennington and others from official enrollment. Cortlandt's father, John William Creed, born in 1801, in Santa Cruz, is first mentioned in the New Haven records when he was accepted as a member of the Center Church of New Haven in 1820. John Creed worked as a waiter and janitor at Yale College. He provided commencement dinners for alumni between 1822 to 1864. During his time at Yale, John was acquainted with Cortlandt Van Rensselaer, a Yale College student, class of 1827. Cortlandt Van Rensselaer was a member of a very wealthy and powerful family. His grandfather, Stephanus Van Rensselaer, class of 1763, was the last patroon of Rensselaerwyck, New York and his father, Stephen, was Lieutenant Governor of New York, 1795–1801.

After college, Cortlandt Van Rensselaer studied law at Yale and completed his law degree in Albany, New York, his hometown. After passing the bar, he studied theology at Princeton and devoted most of his adult live to service in the Presbyterian Church ministering to the free and enslaved backs of Virginia until, "... the excitement on the subject of slavery had so much increased, that he was forced to abandon what had become to him a most exciting field of undertaking." Though not an abolitionist, Cortlandt Van Rensselaer was a member of the American Colonization society, a group he strongly supported. However, the society fell out of favor among black people, people he counted among his true friends, and Cortlandt was discouraged. In a letter to his friend and fellow colonizationist, The Reverend Leonard Bacon, he wrote:

> I delivered an address on 4th July. I have, in so doing, been
> the means of arranging [?] against me, for a time, I fear, my

old friends, the free people of color. I am very sorry for it, but I am not quite in despair yet about the free blacks. I think they must see, in time, that their true friends are the Colonizationist after all.

Thus, there are reasons to suspect that John Creed was acquainted with Cortlandt Van Rensselaer when he was a student at The College, and that they may have been friends. From an inauspicious beginning in New Haven, an immigrant teenager just off the boat from Santa Cruz, John Creed rose to become one of New Havens most successful entrepreneurs. In addition to his duties as janitor at Yale, he was employed as a steward for the Caliopean Society. This society was an offshoot of the Linonian Society, a social and debating club at The College. In context: During their 1810 term elections, partisan politics caused the members of The Limonia Society to elect a president from a Northern state. The Southerners in the society, in a move that foreshadowed national events, immediately withdrew from the society and formed their own. John Creed worked for this exclusively Southern society at Yale from 1828 until 1850 when the society folded. During his time with them, John was responsible for the upkeep of the hall and sundry assignments. Every term he would present a bill to the society's treasurer for payment. The fact that he presented a bill speaks to the issue of his education and administrative skills. This annoyed some members of the Southern society.

Of the remaining debts I have paid 50 dollars ..., besides a number of small bills amounting in all to $35.42 cts. Among those that which holds no inconspicuous station & which I feel myself tin duty bound o mention to you because it ought not to be passed over unnoticed is Creeds bill, the servant who attends to the ha. This fellow charges in my opinion entirely too exorbitant for the little work he has to do, especially when we consider that Charles Harris receives only $2. A term for daily making fires in the library, a much more troublesome duty. I would therefore recommend to you to request your next secretary to whom this power has been delegated to engage him or some one else to do it for less. I think it a waste of money when we can get it at a less price & our treasury is so poor.

John Creed married Vashti Duplex, the first black teacher at the Temple Street School and daughter of the church's first clerk, Prince

Duplex. They were married by the Reverend Leonard Bacon. Whatever the nature of the relationship between the two young men during Cortlandt's time at Yale, John Creed was sufficiently grateful that in April 1835, he named his first born Cortlandt Van Rensselaer Creed. John Creed moved his family in 1848 into a home at 45 West Chapel Street, west of York Avenue, a traditionally white section of the city. By the time his son Cortlandt began studying medicine a Yale, the family lived comfortably. At his death in 1864, John Creed owned several properties. Appointments in his home included mahogany furniture and gold tableware. His estate was valued at $13,468. It is a matter of record that he was versed in business and its administration. The education required for the skills he demonstrated would have been difficult for a black youth to acquire since there were no schools in Connecticut that could have accepted him. Such an education was most often acquired in the form of private tutelage by white benefactors. Cortlandt Van Rensselaer was likely this person for John Creed.

Cortlandt Van Rensselaer Creed was well born into a prosperous, very educated family, prominent and wealthy by African American standards, a family well connected with people in high places. He attended the Lancasterian School, a public school located on Orange and Wall Streets. The school evolved into Hillhouse High School. It is remarkable that Cortlandt Creed was permitted to study at Lancasterian, for it is believed to have had a tradition of not admitting African Americans.

Cortlandt Creed entered the Yale Medical Institute in the fall of 1854 without protest from faculty and students, no record of discontent and discord. His admission to Yale followed the admission of Martin Delany, Daniel Laing and Isaac Snowden, the first African Americans to enroll at the Harvard Medical School, 1850. Their presence was greeted with student great protestations and complaints. Some students would not be identified with them as their fellows

> "... blacks whose company we would not keep in the streets, and whose society as associates we would not tolerate in our houses."

Although the Harvard faculty supported the presence of these three black men and allowed them to attend classes in the beginning, they voted just one day after Christmas of 1850 that these students not be admitted and barred their continuation. After the Civil War, Edwin Howard was admitted to Harvard Medical School and he became her first black graduate in 1869.

Cortlandt Creed took the courses and year of clinical training required during his medical studies at Yale. For his thesis, "On The Blood," he examined the chemistry and physiology of blood in health and disease. He took his oral examinations on January 15 and January 16, 1857 before President Woolsey, representatives of the Connecticut Medical Society, and the faculty of the Medical Institute. Thus, after 2 years and about 4 months and two weeks of formal study, Cortlandt Van Rensselaer Creed and the ten other members of his class received their M.D. degrees on January 16, 1857 from the hands of President Woolsey in The College Chapel. Edward A. Bouchet would have been 4 years and four months old at the time. The Bouchet and Creed, the elders both worked contemporaneously at Yale, both families were members of the Congregational Church and both were involved in the early education of black children in New Haven. By the time Creed died in 1900, nine more African Americans had earned the M.D. from Yale.

Before Creed and Before Bouchet

Before the Hopkins School, before Yale, before the War of Independence, before the incorporation of the city of New Haven, before the *Amistad* zigzagged in the Long Island Sound, before Hillhouse was a student at Yale and before Woolsey was her President, before Creed earned his M.D., before the Civil War, blacks understood the high purpose of education, its personal and practical rewards and the significance of learning in human society. Excluded from the intimate company of scholars, then and even now, blacks sought at great personal risk to learn, and on occasion their courage, energy and ambitions were rewarded with success, isolation and/or punishment even unto death – by law. Without access to established scholarly communities, without access to the intellectual cannons and tradition, under the most challenging circumstances, American blacks sought higher education before and after Bouchet's time at Yale. Blacks knew that for serious study and intellectual achievement, the company of scholars and attendant resources were essential.

Notable among those schools open to blacks during the antebellum period was the Gimore School in Cincinnati. Founded in 1844 by the Reverend Hiram S. Gilmore, the school was large and well equipped. The school prepared students for college, and a fair proportion of them went from there to Oberlin and such institutions that drew no color line for matriculation. Its classical curriculum attracted blacks from throughout the nation, including a sizable group of mulatto children of southern planters. A similar black school was established in 1856 at Tawawa Springs, Ohio, a favorite summer resort of wealthy white Southerners. Sponsored by the Methodist Episcopal Church and controlled almost entirely by whites, the institution became known as Wilberforce University. The first students were largely the children of southern and southwestern planters. The Civil War curtailed southern patronage, and the school passed into the hands of the African Methodist Episcopal Church in 1863. Oberlin, also in Ohio, attracted a sizable contingent of black students from the South as well as the North. From its establishment in 1833 until the Civil War, blacks constituted about 5 percent of the student body. Before the war a substantial portion of them enrolled in the preparatory department. Although the colleges role in the abolition movement is celebrated, black students at Oberlin were not treated the same as whites. Nevertheless, the atmosphere was probably more free of prejudice than at virtually any other predominantly white institution in the country. Many blacks who achieved prominence were enrolled at Oberlin, either as preparatory or collegiate students. Throughout the North and Midwest, blacks acquired education either in racially mixed institutions or in schools established specifically for blacks.

As far back as the eighteenth century, various religious and philanthropic agencies, such as the Society for the Propagation of the Gospel, created separate schools for blacks, Other examples of white benevolence appeared in the form of charity schools and Sabbath schools for blacks. Northern blacks themselves played a significant role in the establishment of schools. Boston opened its first public school for blacks in 1820 and New York followed Boston 20 years later.

In spite of the existence of numerous black institutions that were called colleges and universities, most in fact scarcely qualified as solid secondary schools. Among those that did offer high-quality collegiate education were several institutions established by the American Missionary Association and the Congregational Church, such as Fisk University in Nashville, Tennessee. These institutions, as well as Wilberforce in Ohio and Presbyterian-related Lincoln University in Pennsylvania, had racially mixed faculties and attracted the children of the best colored families where "choice youth of the race" could "assimilate the principles of culture and hand them down to the masses below."

Knowing the odds and responding to chance, hoping daily for new occasion, black Americans came chasing wonderful dreams as only the young can do. A few entered Yale in cooperation with people of good will. While remaining a part of the powerful culture of learning and service they brought with them and refined, African American students tolerated peculiar arrangements at Yale and other colleges in the eighteenth, nineteenth and even into the twentieth centuries; they finished their courses of study without pause and moved on with wonder to witness possibilities. Some even departed with degrees and abiding friendships.

Bouchet and Yale

While in the tercentennial year of Yale, we laud and magnify the practical and intellectual contributions of faculty and alumni, it is especially important to note Bouchet's achievements and those who came before him, those who provide a record of antebellum African-American education, promoted its constructive results and challenged the myth that the education and enlightenment of the race threatened society. They counter the view that African American were neither interested in education nor qualified to educate themselves.

In 1870, Edward A. Bouchet's classmates at Hopkins received more letters of admission to Yale than any other preparatory school in the country. While he could have benefited from the company of blacks he might have known at Fisk, Howard and other historically black schools, Bouchet truly prepared to study at Yale and entered The College on his 18th birthday with the prestige of having graduated from Hopkins Grammar School as valedictorian. He must have known Creed, the first African American to officially enroll and earn a MD (1857) at Yale. They attended the same church. Dr. Creed's mother was his first teacher. However, the distinction of being the first African American to enroll in Yale College belongs to Edward A. Bouchet.

Chapter 2

Edward A. Bouchet
The Years at Hopkins Grammar
School: 1868–1870

Thomas Rodd, Jr.*

> ... somewhat resided ... in these men ... an expectation that
> outran all their performance. The largest part of their power
> was latent. This is what we call Character,—a reserved force
> which acts directly by presence, and without means. It is
> conceived of as a certain undemonstrable force, a Familiar or
> Genius, by whose impulses the man is guided but whose coun-
> sels he cannot impart; which is company for him, so that such
> men are often solitary, or if they chance to be social, do not
> need society.[1]
> –R.W. Emerson, "Character" 1844

Edward Bouchet graduated as valedictorian of the class or 1870
at Hopkins Grammar School, a New Haven secondary school then
located a short distance from the Yale Campus. After Hopkins,
Bouchet, like most of his classmates, matriculated at Yale, where he
again rose to the top of his class. As a Yale undergraduate Bouchet
earned a Phi Beta Kappa key, and as a graduate student he earned
one of this nation's first doctorates, a PhD in the nascent field of
physics. At Hopkins and at Yale, Bouchet stood out both intellectu-
ally and as the sole African American.

Upon entrance into Hopkins Grammar School, Bouchet's life became linked to the school's name and vision, and appropriately so,
since the school's benefactor and namesake, Edward Hopkins, envi-

*8 Sunset Lane, Bradford, NH 03221 (rodd@iamnow.net).

sioned a school for "hopeful youths," by which he meant students causing or inspiring hope in their elders. Unfortunately, what is known about Bouchet's school exceeds what is known about Bouchet himself. Few primary documents related to Bouchet's life have survived. About the only lens through which to view Bouchet during his two years at Hopkins (1868–70) is therefore Hopkins itself.

Hopkins Grammar School took its name from Edward Hopkins (1600–57), the first governor of the Colony of Connecticut, who upon his death in 1657 left his American estates "unto the public use . . . for the Breeding up of hopeful Youths Both at the Grammar Schoole and Colledge for the publique service of the Country in future tymes."[2] Governor Hopkins's bequest led to the establishment of three colonial grammar schools and supported America's only existing college, Harvard.

Governor Hopkins' bequest became this nation's first and oldest charitable trust. Of the schools established through that trust, only two have survived, and the older of these two, Hopkins Grammar School in New Haven, has been in continuous operation since 1660. In 1990, the school, by then a private coeducational country day school located on the western rim of the city, shortened its name to "Hopkins School."[3]

In very crude terms the evolution of American secondary education can be divided into three overlapping but distinct stages: the age of grammar schools, the age of academies, and the age of the public school. Grammar schools characterized the colonial period; academies gained prominence between the Revolution and the Civil War; and public high schools rose to ascendancy after the Civil War. During Bouchet's boyhood, all three types of secondary schools were in existence. Grammar schools were all but extinct, academies were strong, and public high schools were on the rise.

The point to be made about grammar schools as opposed to academies and public high schools is that historically, of the three, only grammar schools devoted themselves exclusively to college preparation. Only grammar schools paid strict homage to the college curriculum. Grammar schools began as the college preparatory schools of the English Renaissance, and since higher education of that time required a heavy, almost exclusive emphasis on classical languages, the "grammar" referred to was predominantly Latin, but also Greek and sometimes Hebrew.

If one remembers that in this country, until the middle of the 18th century, colonial colleges such as Yale and Harvard enforced regulations forbidding student use of English within the college precincts, and that until 1774, as an example, the Laws of Yale were published in Latin, one gets a sense of the rarefied atmosphere intended by institutions like Yale and Hopkins. By design and tradition, to attend Hopkins and Yale as Bouchet did was to be tracked with the country's academic elite.

As a colonial grammar school, Hopkins transplanted and extended a heritage of old world exclusivity. Grammar schools served a public purpose through academic benefit to the best and the brightest. They sought to prepare young men for the rigors of Latin, Greek, and Hebrew, and through that regimen to foster and preserve an educated class from which to draw ministers and other professionals. The reason for this exclusivity had to do with with first colonists' anxiety about religious and cultural self-preservation. The clergy among the early settlers had been educated in England. Replacing them with a new generation of well-trained and theologically correct ministers was an urgent concern—hence Harvard, Yale, and the grammar schools which supported them, Boston Latin, Roxbury Latin, Hopkins Grammar, and others.

At its founding in 1660, Hopkins served to prepare young men for Harvard (then the only existing colonial college). After the founding of Yale in 1701, however, Hopkins students were convenienced by the existence of a more local and (in the view of many) puritanically correct college. As it happened, Yale's first president, Abraham Pierson (HGS 1664),[4] prepared for Harvard at Hopkins, and the story goes that Yale was founded in 1701 so that other Hopkins graduates might be spared the trek to Cambridge, the journey being troublesome and the result (theologically speaking) questionable.

A second point to be made about Hopkins is that, in addition to its focus on college preparation, until the late 19th century, the school served as New Haven's "public" college preparatory school. Until after the Civil War few ideological lines were drawn between public and private schooling. The English ideal of service to the "commonweal" prevailed. Moreover, despite their public responsibility and a measure of public funding, grammar schools remained privately controlled. By today's standards, they held an ambiguous middle ground, and their anomalous state contributed to financial problems. New Haven was lucky. Thanks to the endowment bequeathed by Edward Hopkins, it wasn't until 1796 that the school charged tuition, and after that, again thanks to the endowment, Hopkins set its tuitions well below those of competing academies, most of whom instituted classical courses designed to prepare students for college.

By the time Bouchet entered Hopkins in 1868, New Haven's public schools were growing in strength and purpose, though not in a direction which would soon challenge Hopkins. The New Haven High School began in 1852. The town fathers, however, cast that enterprise in a light different from the one which inspired Edward Hopkins. Edward Hopkins envisioned schools open to the public but narrow in purpose. The New Haven town fathers envisioned schools both open to the public and broad in purpose and outcome, serving children from all sectors. By that thinking, the New Haven High School became an extension upward from city's common schools rather than downward from a college such as Yale. In their search for what would prove to be an elusive, even chimerical, balance between service to the few and service to the many, the New Haven school officials leaned toward egalitarian rather than elitist principles. Forced to choose, the New Haven High School favored equity over excellence, while Hopkins clung steadfastly and precariously to the original intentions of Edward Hopkins.[5]

With limited resources and shaky political underpinnings, the New Haven town fathers struggled valiantly to balance the competing needs of equity and excellence. The former called for curricular latitude; the latter required a specialized and therefore expensive classical curriculum. In the struggles over resource allocation, the question became whether the High School should serve the public broadly defined or that stratum of the student populace who by talent and motivation might qualify for college. To attempt the latter would require funding a classical department, an enterprise which the High School managed to attempt in 1862 when Edward Bouchet was ten years old.

Unfortunately, the 1862 foray into the domain occupied by Hopkins Grammar School resulted in backlash accusations of extravagance, and onto the bandwagon jumped those who from the outset had viewed the High School as superfluous and extravagant, including some members of the Board of Education itself. By 1866, the issue had become whether to fund a public high school at all, and the matter was put to a city wide vote. In the plebiscite that followed, the baby was about to follow the bath water. Happily, supporters of public education carried the day, and the High School survived, though in a weakened state.

As one response to the turmoil, the city fathers sought to skin the cat another way. They entered into negotiations with the Hopkins Committee of Trustees to merge the two schools, proposing to assign the equity side of educational equation to the High School and the excellence side to Hopkins. Their hope was to secure Hopkins' assets as permanent funding for a classical division within the High School, a stratagem which had resulted in the earlier disappearance of Hopkins' sister school in Hartford. Negotiations between the two New Haven schools took place in 1865 and again in 1885 and 1887. For reasons which require no detail here, they came to naught.

In the meantime, the High School found itself overextended in struggling to achieve both equity and excellence. As a consequence, in 1867, when Bouchet was thirteen, the High School abandoned the teaching of Greek, and three years later, when Bouchet was sixteen, they dropped Latin as well. It wasn't until 1877 that the High School reinstituted classical studies, and during that hiatus, New Haven students such as Bouchet found themselves without publicly funded college preparation. Between 1867 and 1877 Hopkins stood practically alone and certainly preeminent as the city school supporting excellence in college preparation.

In 1867 Daniel Coit Gilman, chair of the New Haven Committee on Schools, spun a virtue out of a necessity in his assessment of the situation:

> ... it is but just for us to consider the Hopkins Grammar School to be in reality a Public Latin School... It seems to us that the special preparation of New Haven boys for college ... may be left to the Hopkins Public Grammar School, well

known to be one of the very best schools in the land, and to private schools. The strength of the High School can then be given to the training of a much larger number of boys who do not intend to go to college, but who need special preparation for business pursuits and for further scientific rather than linguistic studies ...[6]

At this point, we enter murky waters. Clearly, Bouchet's arrival at the doors of Hopkins Grammar School in 1868 meant that he was tracked for college. On the other hand, how he got to the school's doorstep is a mystery, a matter for speculation and extrapolation.

According to information handed down, Bouchet's father, William Francis Bouchet, traveled north from Charleston, South Carolina, in 1824 as the servant of John Brownlee Robertson, a Yale undergraduate. After Yale, Robertson settled in New Haven and sent his two sons, both close in age to Edward Bouchet, to Hopkins Grammar School. One of those two sons, Heaton (HGS 1868) preceded Edward Bouchet by two years at both Hopkins and Yale. Heaton's younger brother John, who did not attend Yale, graduated from Hopkins in 1869. Bouchet, of course, was a member of the Hopkins class of 1870.

To characterize the 1860's relationship between the Robertson and Bouchet families forty years after their arrival together in New Haven is not possible. Their relationship was just as likely to have attenuated as it was to have endured and evolved. To assume, therefore, in the interests of sentiment, that the Robertson family sponsored Edward Bouchet at Hopkins is, at the same time, both reasonable and speculative.

Bouchet's family lived on Bradley Street in a section of New Haven which has since yielded to a massive interstate. Bouchet's mother was a northerner, and since she was born in 1818, six years before her husband reportedly arrived in New Haven as a young man, she may have been somewhat younger than her husband. Little has come down to us about Bouchet's father save that he was deemed a pillar of the old Temple Church and that, like his son, he was considered "a man of exceptional character and force of personality."[7]

So far as we know, Edward Bouchet received his early education at Sally Wilson's School, a local "dame school" for the children of New Haven's African American residents. From there he went on to two years at the New Haven High School (1866–1868).[8] Bouchet's reasons for transferring to Hopkins Grammar School can only be presumed. Clearly, he was one of those who by talent and motivation was suited for college. Moreover, as has been noted above, once the High School abandoned its teaching of the classics, Bouchet had few choices in preparing himself for Yale or any other elite college. Whether the Robertson family took a hand in steering the son of their old family retainer towards Hopkins and Yale, or whether Hopkins' low tuition (at $60 a year about 12% of the estimated annual wage of an unskilled laborer)[9] allowed Bouchet's family to support him is unknown. All that can be said is that, absent a viable classical curriculum in the High School, Bouchet's choices in New Haven were Hopkins, another private school, or a private tutor.

Before entering the precincts of Bouchet's Hopkins, some background to his schooling is worth mentioning, issues such as the governance of the school, the expectations for students in general, and the climate for minorities in particular.

Bouchet was not the first African American to attend Hopkins. Amos Beman graduated from Hopkins in 1856. The son of a famous minister at the Temple Street Church (Bouchet's church), Beman might have been the first African American to matriculate at Yale had he not died the year of his graduation from Hopkins.[10] Other African Americans appear in undated photographs of early Hopkins football teams, but more than likely they attended in the 1880's and 1890's when the school became more organized about sports.

Entering Hopkins shortly after the Civil War, Bouchet must have found himself in an environment, if not awkward, then surely indifferent. How that played out in his daily life is hard to say, but not hard to imagine. No matter what, he must have felt like an outsider. In the 1912 Biographical Record of the Yale Class of 1874 he cites as long time friends two of his Hopkins classmates, George Dickerman (HGS 1870) and Henry Farnam (HGS 1870). Since Dickerman and Farnam were also Bouchet's Yale classmates, and since both remained in New Haven, where Bouchet's mother also remained until her death in 1920, it is difficult to know exactly which circumstance accounts for their friendship. It would be pleasing to believe that Bouchet made lasting friends at Hopkins, but that sentiment stands apart from evidence. On the other hand, that Bouchet was poised and adaptive in a white world seems evident from the fact that he not only chose to go on to Yale College and Graduate School, but that he also submitted regularly to the biographical records published by his Yale class.

Despite the predominance of white protestants at Hopkins, over the years the school achieved a remarkable record of accessibility. Successive waves of immigrants sent some of their best and brightest to the school. During the 19th century small numbers of German Jews, Irish, and Italians joined the ranks of the school's "hopeful youths." Prominent among them, James Pigott (HGS 1874) became Connecticut's first Irish Catholic congressman. As an institution expressly dedicated to serving the best and brightest among New Haven's youth, Hopkins had a decidedly liberal cast.

Moreover, two of the Hopkins trustees during Bouchet's years, Alexander C. Twining (HGS 1816) and Theodore Dwight Woolsey (HGS 1816), were noted for their abolitionist sympathies. At the time Woolsey was also President of Yale. Another abolitionist of note had gone off the Hopkins Board five years previous to Bouchet's arrival, having served for thirty-four years. Roger Sherman Baldwin (HGS 1807) acted as chief defense attorney for the Amistad slaves in the 1839 legalities which culminated in one of this country's most famous slavery trials.[11] Baldwin was also a former United States Representative, Senator, and Governor of Connecticut. His grandson Simeon Baldwin (HGS 1857), who also became a famously broadminded jurist and governor, succeeded his grandfather on the Hopkins Board in 1869, when Bouchet was a junior at Hopkins.

These Hopkins trustees were men of obvious clout. The behavior of boys in groups being beyond the pale of strict adult control, no one can say for sure how Bouchet was made to feel as a Hopkins student, but certainly from today's perspective the school's orientation as represented by some of the adults prominently associated with its governance was "hopeful" in the sense that Edward Hopkins had intended the word.

Brief note should here be taken of another aspect of Hopkins governance, one which reflects less well on the Hopkins trustees. The Hopkins Committee of Trustees has been in continuous existence since 1657, and despite the Hopkins endowment, the school has lived most of those years precariously. In the foreground of these struggles and to the great detriment of the school, visionary Rectors (headmasters) alternated with visionary trustees. Over the years, the school seemed incapable of assembling its best talents together in the same room at the same time.

When Bouchet entered Hopkins in 1868, he picked a good time. The school was strong and stable thanks to the exemplary leadership of the Rector preceding Bouchet's years. The Reverend James Morris Whiton, who served as Rector from 1854 to 1864, was a visionary. Even from a vantage point today, he emerges as one of those consummate classroom geniuses equipped to elevate a school to greatness. He began his Hopkins career with an enrollment of six students, and despite some roller coaster years set a course whose momentum carried the school to an enrollment of over two hundred a few years after he left. Whiton liberalized the stodgy classical curriculum, leavening it with English studies, and instituting, among other innovations, a Junior Department, report cards for parents, and public exhibitions of student oratorical skills. Supporting these innovations was Whiton's rare ability to create a school climate so positive that it succeeded in generating student enthusiasm for the rigors of his academic expectations and eliminated all but minor disciplinary issues.

The Hopkins Committee of Trustees, on the other hand, despite their luster, remained close-minded and tight-fisted. Eight in number during Bouchet's years, they included (allowing for some overlap in roles) four Yale professors (including the President, as mentioned above), five alumni, and two parents. Since three of the five alumni trustees were more than fifty years out of Hopkins, the conservatism of their educational views came naturally. These were a tough group of men for Whiton to work with. They niggled about money, curriculum, and equipment, and even went so far as to oppose another of Whiton's farsighted initiatives, his attempt to create alumni support and add to the school's diminished endowment. In 1864 Whiton left exhausted and defeated, a loss to the school of major proportions.

Expectations for students entering Hopkins Grammar School were twofold: expectations while at school and expectations thereafter.

Absent today's standardized testing industry, admissions standards at 19th century Hopkins blended the practical and the intuitive. To enter the Junior class, as Bouchet did, required demonstrated abilities in Latin and Greek. For students entering lower grades, however, admission probably relied on testimonials pointing to such elements of character as industry, deportment, and rectitude. Students were admitted to Hopkins if their parents could pay and if they had good "rec's" from adults in a position to know something about their behavior, if not their piety.

As a result, Hopkins was an academic crucible for the unskilled or unwary. No talk here of a "student-centered" or "life adjustment" curriculum. The school was decidedly teacher-centered and goal-oriented. The school assumed no explicit responsibility for the academic success of its students. Such matters were left to them, to their industry and capacities. Whatever implicit responsibility the school assumed was lodged in the benevolence of teachers.

By today's standards the Hopkins attrition rate was high. Of the ninety students who entered Hopkins between 1854 and 1857, twenty-nine flunked out, a thirty per cent attrition rate. During Bouchet's years, when the school was larger and stronger, the rate was even higher. Of the forty-six students originally in his class, only twenty-nine graduated. Hopkins was not a place for the academically faint of heart or weak of aptitude.

Expectations for Hopkins graduates were also less uniform than one might expect. Even though the Hopkins classical curriculum anticipated years of college to follow, this expectation was honored in the breach. Since early in the 19th century the school had adapted begrudgingly to competition from the academies. As a result, the Hopkins curriculum, while dominated by the classics, evolved over the years into something more versatile and inclusive. Far from being anachronisms, therefore, Hopkins graduates were as well trained to enter commerce as they were to matriculate at college.

Of the twenty-nine graduating members of the Hopkins class of 1870, twenty went on to Yale. Of the eight who terminated their formal education at that point, not much is known, though certainly one of them, Samuel Mallet, went into a family business, the Mallet Hardware Company in New Haven.

The single remaining member of the class of 1870, Edward Dwight, did what seems unthinkable now. He proceeded from Hopkins to Yale

Medical School without benefit of an undergraduate education or degree. In fact, Dwight's traverse was common in the years before university graduate education became monopolistic. At that time law and medicine had one foot planted in graduate school training and the other in apprenticeships. Many Hopkins graduates had already followed the path of Edward Dwight from Hopkins to Yale Medical School, including the brilliant but unbalanced William Chester Minor (HGS 1849), one of the protagonists in Simon Winchester's 1998 bestseller, The Professor and the Madman.[12]

Today's linkage between college and career was years in the future, making Bouchet's journey to college and graduate school all the more distinctive. At that time, even a high school diploma was a rare commodity. High school enrollment accounted for well below two percent of public school enrollment, and of those only a quarter are estimated to have graduated. Moreover, of young men aged 18 to 21 fewer than three percent went to college. Bouchet was exceptional, and not just for reasons of race.[13]

Of the Bouchet classmates who were graduated from Yale, thirteen went on to earn advanced degrees. Seven became lawyers, three earned their PhD's, two became doctors, and one became a minister. Two of the three who earned their PhD's, Edward Bouchet and David Kennedy, did so at Yale and went on to careers in secondary school teaching. The third, Henry Farnam, earned his doctorate in Germany and returned to New Haven to serve as professor of Political Economy at Yale and as Hopkins Trustee for thirty-three years.

Edward Bouchet was schooled in a stately Greek revival structure at the third of the school's succession of five downtown locations, the first having been the New Haven Green. Now the site of the Yale Law School, the Hopkins building attended by Bouchet at High and Wall Streets was built in 1838 for an estimated $2500. It was a one story stucco structure with a half story basement constructed of natural stone. Atop the roof, a handsome cupola housed "a hoarse throated bell which was rung at the change of classes." The large drafty School Room was heated with a coal stove of sufficient proportions to block the Rector's view of students in the back of the room. The upper windows were flanked with blinds, and the basement windows were protected by heavy shutters.

To enter the precincts of Hopkins Grammar School in 1868 was to enter an educational world quite different from the world we know today. Schooling took place all in one room. Classrooms as we know them did not exist. Hopkins was, in effect, a one room school house. Students were summoned in groups to the front of the School Room for purposes of instruction and recitation while the rest of the study body went about their business of studying in preparation for their turns before the faculty. The curriculum was also decidedly different from what we would know today. Though moderated somewhat by that time, the classical curriculum remained in ascendancy, primarily because learning at Hopkins was "for Yale's sake," not its own, and Yale was still a place where "a medieval school man would at once have known where he was."[14]

While easy to disparage, the classical curriculum was rooted in a heritage whose intention, however foreign to today's sensibilities, served long and well in achieving its expressed purpose of training young men (rarely women, needless to say) in eloquence. In the classical sense, eloquence comprised an impressive configuration of virtues—acute critical intelligence, broad intellectual background, diverse experience, and moral prowess—all culminating in a persuasive virtuosity aligned with service to human affairs. Cicero was the reigning deity of classical studies, a model for every student's capacious intellect, virtuoso self-expression, and personal integrity.

Since the Renaissance, the curricular vehicle serving this purpose had been a dead language. The linguistic paragon for the Renaissance was a literary language which no longer existed in vernacular form and which therefore could be learned only in school. Latin was the platform language for educated man of the Renaissance. Not only was the speaking and writing of Latin essential for the affairs of church, diplomacy, and scholarship, but the study of Latin literature provided imitative models for educated men called upon to compose in their native tongue. To learn modern English by learning ancient Latin remains a paradox, but a paradox long honored by tradition and success.

In 1868, Bouchet was therefore entering a vestigial world, a world whose roots traced all the way back to humanism in ancient Greece. Latin and Greek remained essential to the enterprise, but more through force of habit than through any conviction that these languages would prove central in students' later lives. The classical curriculum was holding its own, though not for long. The seeds of change had first sprouted in the 18th century academies and then in the 19th century public high schools. Beginning in 1862, to accelerate matters, Yale's Sheffield Scientific School became free to Connecticut residents under the provisions of the Morrill Act, and the attraction of Sheffield's decidedly more practical course of study created a market for college preparation more broadly based and less classically centered. In 1870–71, the year following Bouchet's graduation from Hopkins, the school instituted an alternative course of study designed to prepare graduates for Yale's "Sheff." So popular did this two tiered system become, that enrollment in Hopkins soon became equally divided between those preparing for Yale College and those preparing for Sheff.

Despite the climate of change, Hopkins held true to its heritage as an institution dedicated to college preparation. In introducing a scientific course, the school was not abandoning ship so much as adapting to shifting winds beyond its control or ken. Hopkins was afloat a sea change transforming American education, and given Bouchet's later commitment to the study of science, one wonders whether he might have chosen the scientific course had he been a year or two younger and benefited from a choice in his Hopkins course of study. The classical course both at Hopkins and at Yale was far more prestigious, but the scientific course heralded both the nation's and his educational future.

Pedagogically, the classical curriculum experienced by Bouchet was almost synonymous with memorization and the doctrine of imitation. Nowadays, the term "imitation" evokes a host of negative connotations, not the least of which are inferior, "unimaginative," and "slavish." And the term "memorization" likely invites the qualification "mere." To adherents of the kind of classical education Bouchet experienced, however, imitation presupposed that sophisticated language patterns could be internalized through regimented memorization and drill. For students, exercises in translation, declamation, and recitation were relentless. Early on, the linguistic paragons of grammar schools were exclusively Latin and Greek, Cicero being the central model. By Bouchet's time, however, even though English as an academic discipline was a generation in the future, English literary models had become frequent as well.

To illustrate the doctrine of imitation, I draw on a retrospective view of 1860 course work in another New England preparatory school:

> What must strike the [modern] reader most forcibly is the omission of English as a subject of formal study. It was by no means neglected, even though it occupied no place on the schedule of classes and examinations. Compositions were required of every boy at frequent intervals, and declamations also. The declamations were not original but were passages from some celebrated work of prose or poetry, committed to memory and delivered with as much oratorical fervor as the speaker could muster ... Mark Anthony's funeral oration over Caesar, Portia's appeal to Shylock, the peroration of Daniel Webster's Bunker Hill Address were pieces frequently chosen. A reading of the complete work, not merely the committing to memory of the isolated passage, was part of the preparation. So with several declamations to give in the course of a year, a boy did probably about as much reading as would be required now in a course in English. Although there was no formal instruction in English grammar, no class drill or individual drill in rules of syntax, the boys learned to express themselves as clearly and as correctly as do our boys now ...[15]

In 1868, when Bouchet entered the precincts of Hopkins Grammar School, the doctrine of imitation still informed most of what went on. Superimposed onto it, moreover, was what 19th century educators called "mental discipline" or "faculty psychology." A panoply of student skills and aptitudes, it was believed, would result from a daily regimen of mental calisthenics requiring prodigious feats of memorization and retrieval. Hopkins subscribed to beliefs long held by Yale and other colleges: "The two great points to be gained in intellectual culture, are the discipline and the furniture of the mind; expanding its powers and storing it with knowledge."[16] Aside from considerations made for math and English, the "furniture of the mind" which Hopkins attempted to arrange in student minds was a knowledge of the classics and classical history. Knowledge was relatively static, a museum of ancient accomplishments. Listed in the Hopkins' 1870 course catalogue were the following areas of concentrated study for

Hopkins Juniors and Seniors: Virgil Cicero, Xenophon, and Homer; Latin grammar and composition; Greek grammar and composition; Greek and Roman history and geography; algebra and geometry. Moreover, the method for evaluating student mastery of that knowledge was as vestigial as the curriculum itself. Even though written examinations had been introduced to Hopkins in 1856, the milieu remained what it had for centuries, a residually oral culture acclimating itself incrementally to the power of print. Textbooks, pen and ink were all in evidence, but student performance and progress depended almost exclusively on how and what the students spoke.

Recitations lay at the heart of the pedagogy of Bouchet's time. Students were required to demonstrate their learning by standing before the Rector and reciting their lessons responsively. The Rector or one of his assistants would test and probe each member of a class as he stood before him at the front of the School room. The interrogation might require recall of a Latin or Greek paradigm or the ability to translate a sentence or passage from Cicero, sometimes spontaneously. Grades were then recorded in a grade book. If not reciting before the Rector in front of the School Room, students were expected to study their lessons in preparation for a next round of recitations. When disciplinary issues arose, they often occurred during these interludes when students were less directly supervised.

But not always. David Kennedy (HGS 1870), the only other Bouchet classmate to earn a Yale PhD, remembers running afoul of the Rector during a recitation, and his account conveys some of the flavor of those proceedings:

"Buck" Johnson [the Rector] was a nervous, peppery, strict disciplinary man ... We had great respect for him. I remember that once in class, fretted because the boy reciting floundered in answering a question, I ventured behind his back to whisper the answer. Quick as a flash, Johnson shouted, "Who told him?" Looking over our faces, he followed up, "Kennedy, was that you?" When I humbly said, "Yes," he continued, "Take your books and go home." Then he waited until I had gathered my belongings and left the building. I went back the next day with trepidation, but he never afterwards referred to the affair."

David Kennedy's remembrance also reveals aspects of the daily routine which would have been familiar to Bouchet:

> The school house ... was too small to accommodate the hundred and thirty pupils, so the members of the upper two classes came only for recitations and went to their rooms in various parts of the city to study. Pupils from out of town, of whom there was a fair number, were located in certain boarding places under the supervision of the Rector. Often several of us would gather together in a fellow's room nearby between classes to study and then meet at the rail fence a short time before recitation, waiting and chatting until the bell was rung. Those periods proved pleasant, friendly and helpful for an interchange of comments upon the coming lessons.[17]

It would also be helpful to know how Edward Bouchet spent those interludes between recitations. Did he return to his home on Bradley Street, half a mile or so distant, or did he group himself with his classmates? Certainly, Bouchet's acumen would have been an asset in those study groups. He knew his stuff, and students who know their stuff normally find themselves drawn into study groups. Similarly, did Bouchet chitchat with his classmates during the moments before the bell summoned them to recitation, or did he stand apart, the lone African American in a white school? Again, the only extrapolation lies in the fact that in the years after his schooling Bouchet stayed in touch with a few of his classmates at Hopkins and Yale. Again, we retreat to a "maybe."

Educationally, Hopkins provided no shortcuts or divergent paths to Yale admission. The trek was long and hard over the same ground covered by every other school attempting to prepare students for Yale. Since Yale's admission standards were prescribed, the Hopkins curriculum followed suit. Much if not all of Hopkins' day to day regimen was simply taken for granted. As a result, the specifics of the Edward Bouchet's course of study at Hopkins are difficult to detail. Bouchet's course of study must be teased out from internal documents. The school had little reason to publicize its curriculum beyond allusions to standard textbooks and billboard names such as Cicero and Xenophon. Along the same lines, the daily schedule and calendar conformed to expectations elsewhere and made no claims worthy of attention. The chief selling point for a school like Hopkins was its success in training students to pass Yale's entrance examinations. The school lived by results.

The Hopkins academic year ran from mid September to mid July. The diary of Alfred Bacon (HGS 1869)[18] notes that Bouchet's first year (1868-9) began on September 12, 1868 and ended July 16, 1869. The length of the school day, on the other hand, is less easy to pinpoint. Bacon mentions two occasions when the Rector dismissed school at 3:30 in the afternoon, implying that normally it ran later. Other documents cite a six hour school day, suggesting a 9:00 a.m. start with perhaps an hour or more allowed for travel home for lunch. Except for a water fountain installed in 1856, Hopkins could boast of few amenities, least of all a kitchen.

Bouchet spent two years at Hopkins. While the school's classical course comprised five years' study, records indicate that he was among many who came for their last two years of secondary school in order to prepare for the rigors of Yale's entrance examinations. Indeed, a good many students came to Hopkins from across the country. During Bouchet's senior year, the school's enrollment of 136 students included students from six states beyond Connecticut and from India and China as well.

Hopkins had a leg up. The school was situated a stone's throw from the Yale campus, its trustees included important Yale professors, and its faculty were invariably Yale graduates. Henry "Buck" Johnson, the Hopkins Rector during Bouchet's years, was Yale trained, as were James Ryder and William Wood, his two assistants, the latter having recently graduated second in his class from Yale. A year or two with such teachers went a long way towards insuring that preparation for Yale was up to date and thorough.

As one might expect, Hopkins Juniors and Seniors were administered large doses of Latin and Greek. Four times a week they recited in Cicero, Latin grammar, Roman antiquities, Xenophon, and Greek grammar. Each of these recitations consumed about forty minutes. In addition, once a week, they recited in Latin prose, modern history, English literature, and natural philosophy (the 19th century term for general science). In total, Juniors and Seniors recited for about two hours a day. During the remainder of the academic day, they either received formal instruction or were free to study and prepare for the upcoming recitations. During Bouchet's years, as mentioned above in David Kennedy's reminiscence, Juniors and Seniors were permitted to study at home or in the nearby rented rooms of out-of-town classmates.

Recalling those days, David Kennedy also reflects that, "The general run of assembly, study and recitation was normal, day by day, passing pleasantly." Kennedy characterized his Hopkins days as a "haze of pleasant memories." More than likely, Kennedy spoke for less than the full complement of his peers. He was a superior student, graduating second only to Bouchet. For most other students, certainly for many of the seventeen members of the original class of 1870 who never made it through, those daily recitation periods must have evoked their share of anguished and humiliating moments.

In addition to the body of knowledge embedded in the classical curriculum, Hopkins also strove to inculcate background and skills suited to practical affairs. Even though the curricular reign of Latin and Greek remained ascendant, over the years Hopkins had added peripheral courses of the sort more typically found in academies. English, modern history, German, elocution, penmanship, even drawing had made their presence known. The records show, for instance, that shortly before Bouchet arrived, a Yale professor was engaged to bring his microscope to school on certain days. One wonders whether Professor Sanford continued to do so as late as 1868, and whether that exposure to the frontiers of modern knowledge played any part in sparking Bouchet's interest in science.

Another feature of the Hopkins course of study were the school's Public Fridays, days every sixth week when the school repaired to a nearby facility for public declamations. These exercises in oratory attracted members of the public as well as parents, and were received so enthusiastically that accounts of their success appeared in the local press.

Initiated in 1856, Public Fridays showcased original compositions as well as literary works committed to memory. James Harman (HGS 1857), for instance, declaimed "Thanatopsis," an 1817 poem by William Cullan Bryant, while his classmate Simeon Baldwin (HGS 1857) orated an original essay entitled "The Great West." How Bouchet fared on Public Fridays we do not know. On another occasion for public oratory, however, the commencement exercises of 1870, Bouchet composed and orated his own valediction, as befitted the scholar who graduated first in his class. The 1870 commencement program also lists him as a second place winner in the declamation exercises held at the close of that school year.

From today's perspective, the 19th century emphasis on oral presentation seems archaic, if not arcane. In a world now swamped by information in visual form, it is difficult to imagine an education so bottlenecked by oral presentation. These, however, were days well before the emphasis on silent and rapid reading. Academically speaking, 19th century knowledge was boundaried by a small corpus of prescribed texts, and wisdom was officially received from but a handful of those. Edward Bouchet lived in a school world of both sound and print, but more so than students today, his world of sound reverberated with literary cadences and diction culled from a relatively small number of authors and works. At heart, Bouchet's education at Hopkins both trained him in the eye language of ancient languages and enveloped him in the ear language of formal literature. Thus, his sensorium reverberated with rhetorical echoes of the past, imprinting a repository of linguistic templates supportive of proficiency in his own language as refracted through Latin and Greek. This was the doctrine of imitation at work. More was done with less.

Edward Bouchet graduated first in his class at Hopkins. For two years his recitations were graded, and his grades recorded and averaged. It was not the custom of the times to element such averages with subjectivity. Bouchet ranked first in his class because he, above others, had mastered the material. Sentiment played no part. Bouchet was good—very good—and nothing underscored that fact quite so emphatically as his success in this quintessentially harsh male academic environment, picayune and regimented to be sure, but also unyielding in its precision and rigor.

Bouchet's recitation days were not over. He faced more at Yale and before that, the entrance exams at Yale. Hopkins finished in mid July. Short days later, Bouchet crossed High Street to present himself for entrance exams at Yale. These exams were but versions of what he had experienced at Hopkins, successive days during which his ability to translate Cicero and Xenophon ex tempore and parse

Latin and Greek paradigms was put to the test.

While no one can say for sure how fully or rigorously Yale candidates were examined early in the summer of 1870, Yale's published admissions requirements convey what was officially expected of applicants:

> Candidates for admission to the [Yale] Freshman Class are examined in the following books and subjects:
>
> Latin Grammar, including Prosody.
> Sallust–Jugurthine War, or four books of Caesar.
> Cicero-Seven Orations.
> Virgil–the Bucolics, Georgics, and the first books of the Aeneid.
> Arnold's Latin Prose Composition, or the Passive voice
> (first XII chapters).
>
> Greek Grammar–Including Prosody.
> Xenophon–Anabasis, first three books.
> Greek Grammar, including Prosody.
>
> Greek Reader–Jacobs, Colton, or Felton. In place of the
> Greek Reader the candidate is at liberty to
> offer the last four books of Xenophon's Anabasis or
> four books of Homer's Iliad.
>
> Higher Arithmetic–Including the metric system of
> weights and measures.
> Day's Algebra, to Quadratic Equations.
> Playfair's Euclid–first two books (introduced in 1856)
> The first, third, and fourth books of Davies'
> Legendre Elements of Geometry, or of Loomis'
> Elements of Geometry, may be offered as a substitute
> for Playfair's Euclid.
>
> English Grammar and Geography, a thorough knowledge
> of which will be required.[19]

Such were the "discipline and furniture of the mind" that aspirants to Yale hoped to exhibit and for which Hopkins sought to prepare them. In 1870 the number of pages of Latin expected to be mastered by Yale applicants ran to three hundred and twenty-nine, while the number of pages of Greek exceeded two hundred.[20] No doubt, as is often the case with admissions criteria, the bar for Yale applicants was set deliberately and perhaps impossibly high. Even so, given his extraordinary success both at Hopkins and at Yale, Bouchet must have come as close as anyone in his efforts to clear it.

In addition to his academic prowess, Edward Bouchet must have possessed extraordinary resilience and courage. That he was paradigmatic of Edward Hopkins' "hopeful youths," he himself might have recognized. But whether he was self-conscious in his role is another matter. Certainly he was not likely to have known that among the property items which comprised Governor Hopkins' 1657 bequest was a single slave—"the Negar," as the unfortunate was inventoried.[21] Even so, as the descendent of slaves amidst the whiteness of Hopkins, Bouchet could hardly have ignored the ironies of his context. He learned well from his father, it seems, "a true, old-fashioned gentleman," but beneath that admirable stature and stoicism, he must have entertained moments of recognition that, in personal terms, all was not easy.

I close this chapter on a personal note. I came to Hopkins in 1971 as a thirty-year-old English teacher and coach, and left in 1999. The last ten years of my Hopkins years were spent as Headmaster or (as previously termed) Rector. Over time, as I grew conversant with the school's history, the weight and substance of its heritage, now playing itself out in a fourth century, became an inspirational center and compass. That compass invariably pointed in one direction, however, towards a respect for the school's distinctive individuals. Though hardly unique in its dedication to excellence, Hopkins has achieved a longevity which permits more than the customary institutional boasting. The school's long history spotlights numerous distinguished graduates, scores of them eccentric, some of them outright mad, and so many of them exceptionally bright and accomplished. None, however, inspires institutional pride to the same degree as Edward Bouchet.

As an institution, Hopkins has known and will continue to know many deficiencies, but counted among its many counteractive strengths and accomplishments has been its capacity to cut to the chase and focus directly and intensely on training for the best and brightest among the hopeful youths within its grasp.

The headnote to this chapter quotes briefly from Emerson's famous essay on character. Firmly rooted in the oratorical world experienced by Bouchet, Emerson makes for discursive and difficult reading. It is a shame, because Emerson wrote an enduring monument to human understanding. Certainly, he puts his finger on what Edward Bouchet was all about.

Character can exist apart from deeds, and character apart from deeds can serve as a great model, teacher, and influence. Edward Bouchet was, in Emerson's terms, "a presence ... without means." Look for a record of his accomplishments beyond schooling, and you will be hard put. Conversely, scrutinize his character, and you unveil a universe. Officially, Bouchet said little if anything to us. Off the record, he said everything that ever was of value or worth. Courage, dignity, endurance, integrity, intelligence—it is difficult to imagine a priority of human values which he did not demonstrate, and all under circumstances configured to suppress and devalue. In Emersonian terms, Bouchet was a natural power, like light and heat, an inspirational force liberated from circumstance of time and place.

A reproduction of Edward Bouchet's portrait now hangs in the central hallway of Hopkins School. The original hangs in the nave of Sterling Memorial Library at Yale. Both Hopkins and Yale have learned from Bouchet, though less about him perforce than about themselves. Regret can't rewrite history, but recognition can motivate and inspire. Bouchet is behind us, though not his future. Bouchet inspires not just because he succeeded in the narrow, circumscribed world of 19th century Hopkins and Yale. Bouchet inspires because he was a significant human being, albeit a presence without means. The passage below postdates Bouchet's Hopkins years by more than a century. If we cannot today hear Bouchet, we can hear this young woman:

> Edward Alexander Bouchet's portrait hangs in the hallway of Baldwin Hall at Hopkins Grammar School in New Haven, Connecticut, along with many of the school's past headmasters and distinguished alumni/ae. I walk by his portrait every morning on my way to math class and sometimes I look at him and wonder what his experience at Hopkins was like in comparison to mine. He undoubtedly had to recognize the uniqueness of his presence at Hopkins, not only because he was an African American but because he was also a brilliant student ...
>
> I am now a member of the graduating class of 1993 at Hopkins

> School. As one of the few African Americans who has attended
> Hopkins for six years, I am also in a unique position. I am cap-
> tain of the lacrosse team and the only African American on
> the team. I remain the only African American in most of my
> classes, and while my color is not a primary concern, it is ap-
> parent.
> While there is no documented proof of Edward Bouchet's ex-
> perience at Hopkins, I can imagine his feeling of singularity.
> However, in order for him to have been highly successful at
> Hopkins, his education must have been his main concern ...
> He valued opportunity over color, which has been an inspira-
> tion for my success at Hopkins.[22]

Emerson was on the mark in predicting that "We shall one day see
that the most private is the most public energy, that quality atones
for quantity, and grandeur of character acts in the dark, and succors
them who never saw it."[23]

Bibliographical Note

In writing this chapter, I have relied heavily on two works. For the
history of the relationship between Hopkins and New Haven public
schools, Ben Justice's Yale Senior History Essay provided a wealth
of background and insight. It is a marvelous piece and I am greatly
in his debt. Without the second work, Tom Davis' Chronicles of
Hopkins Grammar School, 1660–1935, no one can even begin to write
about the school. Davis was a history teacher at Hopkins, and his
Chronicles began as compilation of archival material, evolved into a
first rate institutional history, and finished as a PhD dissertation at
Yale.

A third work central to this undertaking has been the Catalogue
of the Trustees, Rectors and Alumni of the Hopkins Grammar School
of New Haven, Connecticut, published in 1902. No author is credited
for this monumental task; it was a group effort. But for historians of
the school, the volume is invaluable because it lists everyone associated with the school for more than two hundred years, their dates of
attendance, employment, or trusteeship, as well as their subsequent
schooling and careers. Information about Bouchet's classmates, for
instance, relies on this catalogue.

Informing this chapter has been my own background in the history
of American education begun as a Klingenstein Fellow at Teachers
College, Columbia. Central to those studies has been a concern for
oral traditions as they played out in the teaching of rhetoric and composition. The classical course, as Bouchet's Hopkins experience was
usually termed, strikes us now as narrowly retrograde, but its ironies
and paradoxes highlight divergent but historically valid assumptions
about language acquisition. An exploration of this topic can be found
in my "Before the Flood: Composition Teaching in America, 1636–
1900," a piece which, as published, unfortunately suffered more than
its fair share of editorial indignities.

Bibliography

Broome, Edwin Cornelius, A Historical and Critical Discussion of
College Admission Requirements. Columbia University Con-
tributions to Philosophy, Psychology and Education, Vol XI,
Nos. 3–4. New York: Macmillan, 1903.
Catalogue of the Trustees, Rectors, Instructors and Alumni of the
Hopkins Grammar School of New Haven, Connecticut. New
Haven: Dorman Lithographing Co. 1902
Davis, Thomas B. Jr. Chronicles of Hopkins Grammar School, 1660–
1935. New Haven: Quinnipiack Press, Inc. 1938.
Justice, Benjamin J. "Good Enough for the Best, Cheap Enough
for the Rest: Nineteenth Century Secondary Education in
New Haven." Yale University Department of History Senior
Essay. Unpublished essay, 1993.
Rodd, Thomas, Jr. "Before the Flood: Composition Teaching in
America, 1636–1900." English Journal 72. 2 (1983): 62–69.
Soloway, Scott M. "The Birth of a Country Day School: A Study of
the Hopkins Grammar School. Unpublished essay, 1984.

Endnotes

1. Ralph Waldo Emerson, "Character" [Essays, Second Series]
in Ralph Waldo Emerson: Essays and Lectures. Ed. Joel
Porte. (New York: The Library of America, 1983) 495.

2. Quoted in Davis, 71.
3. The Hopkins bequest funded grammar schools in New Haven, Hartford, and Hadley, Massachusetts. Hopkins School in New Haven is now a private independent school serving grades 7–12. The Hartford school vanished, having merged with the Hartford Public High School in the middle of the 19th century. Also during the 19th century the Hadley school became first a private academy and then later a privately controlled public high school, a role which it continues to this day. See Margaret Clifford Dwyer, Hopkins Academy and the Hopkins Fund, 1664–1964 (Hadley, Mass: The Trustees of Hopkins Academy, 1964).
4. After proper names (HGS) abbreviates Hopkins Grammar School, and the date following designates the year of graduation from the school.
5. Justice, 26–31. For an excellent but oddly titled historical account of how this interplay between equity and excellence has played itself out in American education, see David F. Labaree, How to Succeed in School Without Really Learning: The Credentials Race in American Education (New Haven: Yale UP, 1997).
6. Annual Reports of the First School Society, 1867. Quoted in Justice, 30–31. Daniel Coit Gilman became a pivotal figure in the history of American higher education. An 1852 Yale graduate, he spearheaded the organization of Yale's Sheffield Scientific School, and then, having failed to succeed Theodore Dwight Woolsey as Yale President, went on to become the founding president of Johns Hopkins, this country's first institution of higher learning to focus on graduate as opposed to collegiate education.
7. Yale University, Biographical Record of the Class of 1874: Part 5, 1909–1919 (New Haven, 1919) 84.
8. Dictionary of American Negro Biography. Ed. Rayford W. Logan and Michael R. Winston (New York: Norton, 1982) 50–51.
9. Justice, 41–42.
10. Soloway, "Birth of a Country Day School," 7.

11. For a full account of Baldwin's role see Howard R. Jones, Mutiny on the Amistad (New York: Oxford UP, 1987).

12. Simon Winchester, The Professor and the Madman; A Tale of Murder, Insanity, and the Making of the Oxford English Dictionary (New York: Harper, 1998). From his quarters in an English hospital for the criminally insane, Minor used his lucid moments and extraordinary linguistic talents to contribute substantively to the compilation of England's greatest dictionary.

13. Labaree, 64–5.

14. George W. Pierson, Yale College: An Educational History, 1871–1921 (New Haven: Yale UP, 1952) 71.

15. Arthur Stanwood Pier, St. Paul's School, 1855–1934 (New York: Scribner's, 1934) 94–5.

16. "The Yale Report of 1828" in Richard Hofstadter and Wilson Smith, American Higher Education: A Documentary History (Chicago: U. of Chicago, 1961) 278. This famous report has stood the test of time in arguing the value of a liberal arts core curriculum, albeit one grounded in the classics.

17. Quoted in Davis, 398–400.

18. Quoted in Davis, 393–8.

19. Quoted in Broome, 49–50.

20. Broome, 66.

21. "An Inventory of Governor Hopkins' Household Equipment and Personal Goods." Quoted in Davis, 592.

22. Keisha Blake, "A Portrait of Edward Bouchet" (Unpublished essay, 1993).

23. Emerson, 508.

Chapter 3

Edward Alexander Bouchet
The Master Teacher and Educator

H. Kenneth Bechtel*

Introduction

On October 28, 1918, a brief notice in the obituary column of the
New Haven Evening Register reported that Edward A. Bouchet had
entered into rest at his late residence, 94 Bradley Street. The following day, another small notice announced that funeral services for
Mr. Bouchet would be held at St. Lukes Episcopal Church with in-
ternment in the family plot at Evergreen Cemetery. Today, an un-
marked depression in the grass under a large oak tree is all that
remains of Edward Bouchet's final resting-place. A forgotten grave,
a forgotten man. Hardly the appropriate testimonial for the man
who at the age of twenty-four became only the sixth person in the
United States to be awarded the Ph.D. in physics.

Edward Bouchet graduated class valedictorian from the prestigious Hopkins Grammar School in New Haven, Connecticut in 1870.
He enrolled at Yale and completed his undergraduate degree in 1874
(Phi Beta Kappa). And, after two years of graduate study, was
awarded the Ph.D. in Physics from Yale in 1876 for a dissertation in
geometrical optics entitled, *On Measuring Refracting Indices.* Such

*Wake Forest University, Department of Sociology, Winston-Salem, NC 27109
(bechtel@wfu.edu).

educational accomplishments should have placed Edward Bouchet among 19th century America's scientific elite. Lesser academic credentials proved advantageous to many of Bouchet's peers who quickly moved into the forefront of American higher education, medicine and science. Yet, the name Edward Alexander Bouchet is never mentioned in discussions of the early founders of American physical science.

With his superior intellectual ability, prep school training, and undergraduate and graduate degrees from one of the most prestigious universities in the country, Bouchet was unable to follow the same path as his equally talented contemporaries. All the honors and academic credentials could not overcome the fact that Edward Alexander Bouchet was Afro-American. For Bouchet, the ticket of higher education held no promise as race prevented him from sharing the rewards bestowed on those of similar ability and educational achievement.

Family and Educational Background

Edward Alexander Bouchet was born on September 15, 1852 in New Haven, Connecticut to William and Susan Bouchet. Edward was the first boy and youngest in the family with three older sisters. New Haven's Afro-American community provided much of the city's unskilled and domestic labor and William's occupation was typical as records list his occupation as "helper," "laborer," or "janitor."[1] Susan was a housewife but literate and amazingly outlived all but one of her children, dying in 1920 at the age of 102.

Many members of New Haven's Afro-American community were involved in reformist and abolitionist activities, which were usually organized around the church. The Bouchet's were no exception as all were members of the Temple Street Congregational Church. The Temple Street church was a stopping point along the Underground Railroad for fugitive slaves and church members were involved in the political and social issues of concern to the Afro-American community.[2] Edward and his father were active in church affairs as William was both a member of the Society's Committee and church clerk, while Edward served as church sexton during the time he was at Yale. The church would remain a prominent focal point for Edward throughout his life.

Education was stressed among a large segment of New Haven's Afro-Americans, and all the Bouchet children attended school. Nevertheless, Afro-Americans in Connecticut had to attend "colored schools" provided by the school board. Edward attended the "Artisan Street Colored School" located five blocks from his home. As were most of the schools for Afro-Americans in the city, it was small (only 30 seats), ungraded, and only had one teacher; Mrs. Sarah Wilson.[3] The influence of this women can only be imagined as she clearly played an important role in nurturing his academic abilities.

In 1868 Edward Bouchet was the first Afro-American to be accepted into Hopkins Grammar School, a preparatory school for the classical and scientific departments at Yale College. The course of study ranged from three to five years in proportion to the age and intellectual maturity of the student. As stated in the term for admission, candidates who wanted to enter at an advanced level had to pass an examination in the studies gone over by that class.[4] Edward entered Hopkins as a member of the second class, indication that he had acquired a excellent education and was already a gifted student. It was at Hopkins that Edward was introduced to classical education. He studied Latin and Greek grammar, geometry, algebra, and Greek history. He graduated in 1870 ranked first academically and class valedictorian.[5]

Edward Bouchet entered Yale in the fall of 1870 and continued to excel academically. When he graduated in 1874, his career grade point average was 3.22 on a 4-point scale ranking him sixth in a class of 124.[6] This exceptional performance resulted in Edward being elected to Phi Beta Kappa. Being the only Afro-American student at Yale, it is certain he was a marginal member of the student community. According to statistics of the class of 1874, only four members of that class did not join any of the secret societies; Edward was one of those four.[7] Outside of Yale, however, Edward Bouchet was being noticed. During Edward's senior year at Yale he had been contacted by Alfred Cope, a member of the Board of Managers for a private high school for Afro-Americans in Philadelphia run by the Society of Friends, the Institute for Colored Youth (ICY). Cope had an abiding interest in the sciences and wanted the students at the Institute to receive instruction in this area. In his efforts to develop a scientific department, Cope sought out Bouchet and offered him a teaching position. He further encouraged Bouchet to remain at Yale and pursue graduate studies in physics. Bouchet responded that he would be unwilling to undertake two years of graduate study unless he could receive financial support and be guaranteed a starting salary of $1500 dollars a year. Cope agreed to this stipulation and proceeded to completely finance Bouchet's graduate education at Yale.[8]

The Yale graduate school had been operating since 1847, primarily offering a MA degree after a residence requirement had been met. In 1860, the decision was made to develop a separate doctoral program leading to the Ph.D. The original minimum requirements included completion of a four year undergraduate program, two years of graduate study in residence, passing a comprehensive exam, knowledge of classical languages, and completion of a dissertation providing evidence of original scholarship.[9] The new program awarded the first doctorates in the United States in 1861 to James Whiton, Eugene Schuyler, and Arthur Wright. Wright was the first to be awarded a doctorate in Physics and would be Bouchet's major professor in graduate school.[10] During his two years in the graduate school, Bouchet paid special attention to chemistry, mineralogy, and experimental physics. In 1876 Edward Bouchet completed his dissertation on a new subject of geometrical optics entitled "On Measuring Refracting Indices" and earned his place in history as the first Afro-American to earn the Ph.D. from an American university.

The Philadelphia Years: 1876–1902

The Institute for Colored Youth (ICY) was a high school for Afro-Americans founded in Philadelphia by the Society of Friends (Quakers). Established as the direct result of a bequest by Philadelphia Quaker, Richard Humphreys, the ICY began as a rural farm school in 1840. Owing to poor planning and financial management, the farm school was disbanded in 1846. Renewing their efforts in the fall of 1849, the Board of Managers established an apprentice program for Afro-American youths in Philadelphia. At the suggestion of Afro-American tradesman, an evening school was also added. By 1852, a day school program had been started and the ICY moved to a permanent building in the heart of Philadelphia's Afro-American community.[11]

During the 1850's and 1860's, Afro-American children were denied admittance to Philadelphia's white schools which, in effect, precluded any possibility of getting a high school education. The ICY, therefore, provided an important educational opportunity by offering free secondary education.[12] The original managers of ICY, such as Alfred Cope, were firm believers in the value of literary education and felt Afro-Americans were capable of unlimited educational achievements. With this educational philosophy, the ICY offered a curriculum that included ancient history, geography, Greek and Latin classics, algebra, geometry, and chemistry.[13]

As the ICY grew under the able leadership of Cope and Fanny Jackson Coppin, the dynamic principal from 1865 to 1902, the school gained an enviable reputation as the premier secondary educational facility for Afro-Americans in the country. And, in an effort to expand the school's capabilities, Alfred Cope in 1874 gave anonymously $40,000 of coupon bonds to establish what he called the Scientific Fund. His intention in creating the fund was to promote learning in the principles of applied science by "every appropriate means" including lectures, laboratories, machinery, and textbooks. Cope even envisioned developing medical instruction at the school in the distant future.[14] It was the establishment of the Scientific Fund that led Cope to contact Edward Bouchet about coming to ICY to head the new science program.

Edward Bouchet arrived in Philadelphia in the fall of 1875 and taught at the ICY for the next twenty-six years. Although Philadelphia at the time was as segregated as any southern city, there existed a supportive environment for a man of Bouchet's abilities. The city's Afro-American population, the largest in the north, had made considerable progress in education during the decades preceding Bouchet's arrival. By 1849, half the city's Afro-American population was active in one or more of the many literary societies established by the Afro-American community. In 1860, there were fifty-six private schools for Afro-Americans in the city. The ICY, the only high school for Afro-Americans, played an important role in training the thousands of Afro-American teachers that were needed throughout the country following emancipation.[15] Having secured an excellent position, both personally and financially, Bouchet must have been optimistic about the future of Afro-American education in general, and the leadership that the ICY could provide in particular.

Bouchet's starting salary had initially be set at $1500 as he had requested in his negotiations with Cope, but financial difficulties at the school resulted in his salary being reduced to $1200 even before he arrived in Philadelphia. The managers of the ICY never granted raises unless requested by individual faculty members, and even if requested they were not always granted. Either Bouchet never requested or if he did make a request he must have been turned down for his salary remained $1200 a year for the next 22 years. In 1898 he had his salary reduced to $1100 where it remained for the rest of his employment at the ICY.[16] In today's dollars Bouchet was earning the equivalent of between $25,000 and $30,000 a year. In addition to his salary, however, Bouchet probably never had to pay rent or make mortgage payments. During the twenty-six years he resided in Philadelphia, he lived at two addresses.[17] From 1876 until 1888 he lived at 821 Bainbridge Street, a house owned by the Mangers and located next door to the ICY. Past principal Ebenezer Bassett and current principal Fanny Jackson Coppin had both lived in this house more or less rent free. Bouchet was probably given the same arrangement.[18] From 1889 until he left Philadelphia in 1902, Bouchet lived at 830 Lombard Street, an apartment building that was also owned by the mangers near the original location of the ICY at 9th and Lombard.

Although we have very little information that could provide a window into Bouchet's daily life in Philadelphia, a few observations can be made. Scattered references in the reports made to the Board of Mangers of the ICY tell us something of his teaching experiences. He was hired as a science teacher, and during his tenure with the ICY, Bouchet taught all the science classes offered at the school. A typical day would have found him teaching five classes a day with a short demonstration laboratory session. And although the courses he taught varied over his 26 years at the school, he primarily offered courses in Chemistry, Physics, Astronomy, Physical Geography and well as Physiology and Entymology. In 1894, for example, he taught a total of 112 students in eight different classes over a two-semester school year. In his 1893 year end report Bouchet stressed the importance of regular attendance and suggested that prizes be awarded as an incentive. However, he made special note of the fact that three students had perfect attendance through out the year without any incentive at all.[19]

Given his training and the courses he taught, it is not surprising that Bouchet would be interested in improving the level of education for his students. One constant issue that occurs on a regular basis is his request that the Managers provide funds and space for a real science laboratory where students could do individual experiments. Bouchet made such a request in 1884, suggesting that such a laboratory could be included in the plans for the new Industrial Building to be built on property adjoining the ICY. Nevertheless, in 1890, Bouchet renewed his request for larger laboratory space as the new Industrial Building, completed in 1889, did not contain the lab facilities he had hoped for.[20] Fanny Jackson Coppin, the principal of the ICY supported Bouchet in his requests for lab space. In her report to the managers in 1890, Coppin asks the Managers to make provisions for individual laboratory experimentation by students in Bouchet's classes. In her pleas, she makes note of the desire for scientific study among ICY students and that these desires should "be encouraged as much as possible."[21] Coppin further noted, that while may of ICY alumni had earned distinction in the various professions, she rarely heard of an ICY student "devoting himself to scientific pursuits."[22]

In addition to his classroom duties as teacher, Bouchet also gave numerous talks on various scientific topics that were offered to all ICY students and faculty. Subjects of these talks ranged from astronomical and mathematical geography to discussions on the metric system of weights and measures.[23] Bouchet took his scientific expertise outside the walls of the ICY as he reportedly was involved in extension work in the local Afro-American community by giving public lectures and demonstrations on various scientific topics. He also was a member of the Franklin Institute, a foundation for the promotion of the mechanic arts chartered in 1824. Maintaining his Yale ties also was important for Bouchet, as he was an active member in the local chapter of the Yale Alumni Association, faithfully attending all meetings and annual dinners where he was received with cordiality and respect.[24]

Outside of his scholarly pursuits, evidence clearly indicates that Bouchet was an active participant in the larger Afro-American community in Philadelphia. his active involvement with the church, he joined St. Church, the oldest Afro-American Episcopal church in the country. St. Thomas had a reputation as the home of Philadelphia's Afro-American elite. Du Bois characterized the congregation at St. Thomas as being the "most cultured and wealthiest" of the Afro-American community. Octavus Catto, Bouchet's predecessor at ICY was a member, as was his current principal, Fanny Jackson Coppin.[25] Bouchet was not only a member but he also served on the vestry and acted as their secretary for many years. The church Bishop appointed Bouchet to be a lay reader, which gave him the opportunity to take part in church services.[26]

Roger Lane notes that one of the most striking features of 19th century Philadelphia's Afro-American community was the size and diversity of various clubs, organizations, and associations. According to Lane, the largest of these organizations were the various fraternal orders such as the Masons and Odd Fellows.[27] No evidence currently exists to indicate whether Bouchet was a member of any of these fraternal organizations. As noted previously, he was one of only four members of his class at Yale that did not belong to any of the secret societies on campus. Whether this was by choice or racial exclusion on the part of the societies is not known. Given the scope of the fraternal organizations in Philadelphia, Bouchet would have been aware of their existence. In fact, Bishop Levi Coppin, the husband of the ICY principal Fanny Jackson Coppin, served as the Grand Chaplain of a Mason Lodge.[28]

While Bouchet's involvement with the fraternal organizations is open to question, he was clearly involved with another important Afro-American organization in Philadelphia, the building and loan association. These small economic organizations, operating somewhere between a bank and a club, provided the necessary financial support that allowed many working class Afro-Americans to purchase homes.[29] They also provided financial benefits to their subscribers and, as Lane notes, were attractive investments for the more economically successful members of Philadelphia's Afro-American community.[30] One such organization, the Century Building and Loan Association organized in 1886, listed Edward Bouchet as one of its officers.[31]

It is unfortunate that so little material about Bouchet's twenty-six years of residence in Philadelphia is available. But, the few pieces of information that do exist reveal that Edward Bouchet was more than a passing spectator to the events shaping the Afro-American community in Philadelphia during the late 19th century. Base on occupation, income, and social connections established through church and civic involvement, Bouchet was clearly one of the more prominent members of Philadelphia's Afro-American elite.

Edward Bouchet's initial optimism about the future of the ICY and his own personal aspirations would eventually fade. The year Bouchet arrived at the ICY, his benefactor, Alfred Cope, died. Although the schools able principal, Fanny Jackson Coppin, attempted to carry on the ideals he had established, many of the newer managers of the ICY did not share Cope's progressive attitudes and unbiased opinion of Afro-American capabilities.[32] Thus, they made no voluntary effort to expand or improve advanced scientific education at ICY. However, through efforts of Coppin, Bouchet's repeated requests for a laboratory were finally realized in 1891.[33]

But, by 1896, many Philadelphia Quakers were becoming disillusioned with the Afro-American community. Reflecting the changes taking place in the larger society, many "felt that the race was incapable of responding to the efforts put forth on their behalf."[34] The managers began to rely on such Quaker educators as Francis Gummere, a professor at nearby Haverford College, in their planning for the future direction of ICY. In 1894, Gummere suggested that studies at the Institute be simplified, stating that the courses were "pitched too high."[35] The new managers shared this view and were becoming more enthusiastic and receptive to the educational philosophy of Booker T. Washington. By the end of the century, the managers had become openly hostile to classical and academic education, favoring the industrial training of Hampton and Tuskegee. Rather than fight a losing battle, Coppin resigned as principal in 1902. In their efforts to redirect the ICY, the managers proceeded to fire all the teachers inc After twenty-six years of hope Afro-American education, Bouchet's dream was shattered by racism accommodation.

The Troubled Years: 1902–1918

At the age of fifty, Edward Bouchet found himself in the summer of 1902 out of work. In what was to become a common occurrence in the years ahead, Edward Bouchet returned to New Haven and his family home. That September he embarked on the first of a series of moves that would take him across the country in search of employment and teaching positions.

In September 1902 Edward Bouchet boarded a train and traveled 1000 miles to St. Louis, Missouri were he took a job as teacher of math and physics at Sumner High School. Currently known for its recent famous graduates, Dick Gregory and Arthur Ashe, Sumner High School was founded in 1875. The school was named for the prominent anti-slavery senator Charles Sumner, and was the first high school for Afro-Americans west of the Mississippi. The first principal and teachers were white, and with no model to follow established a rather rigorous curriculum. No "special" courses were developed and substandard performance was not allowed. Primary emphasis was placed on courses in English, Latin, mathematics, history and the natural sciences. The uniqueness of Sumner must be underscored given that in 1875 most Afro-Americans did not go to school, let alone a high school offering a high quality college preparatory program.[36]

In 1879, Oscar Waring, a graduate of Oberlin College, became the first Afro-American principal of Sumner. From the very beginning of his term, he chose the best Afro-American instructors in the country to teach standard high school courses at Sumner.[37] It was Waring who hired Bouchet to begin teaching at Sumner in the fall of 1902. And, it is likely that Fanny Jackson Coppin played a role in securing this position for Bouchet as she also was a graduate of Oberlin and probably had reason to become acquainted with Waring through her efforts on behalf of education for Afro-Americans. Bouchet's tenure at Sumner High School was brief as he left the school after only fourteen months in November of 1903. As an interesting side note, if Bouchet had remained at Sumner he would have been joined by another brilliant Afro-American scientist, Dr. Charles H. Turner. Turner, famous for his pioneering work with insects, received his doctorate from the University of Chicago in 1907. He joined Sumner High School in November 1908 and remained there until his death in 1923.[38]

Upon leaving Sumner, Bouchet spent the next three years engaged in non-academic work. In November 1903, he took a position as the business manager and assistant superintendent for the Provident Hospital in St. Louis.[39] A private facility, Provident was the "colored" hospital in segregated St. Louis. No doubt Bouchet's science background combined with his financial knowledge gained from serving on the board of directors for a savings and loan in Philadelphia served him well in securing this position. Nevertheless, Bouchet's employment with the hospital was brief as after only seven months of working for Provident, he quit in May of 1904 and became a United States Inspector of Customs at the Louisiana Purchase Exposition in St. Louis. Bouchet remained in this capacity for almost a year, leaving in March of 1905.[40] In a biographical sketch for a Yale publication, Bouchet remarked that he was able to obtain the customs position through the offices of former United States Representative Charles F. Joy.[41] Joy was no stranger to Bouchet, however, as they had been classmates at Yale. Upon leaving his position with the customs office, Bouchet returned home to New Haven in March, 1904.

Altogether, Bouchet lived in St. Louis for only three and half years. And, as was the case in Philadelphia, little direct evidence of his short stay remains that might provide details of his life and activities. Nonetheless, indirect evidence and information provides the basis for some historical speculation.

From September, 1902 until leaving St. Louis in May, 1904, Bouchet lived at 2724 Morgan Street. There is no evidence to indicate whether this residence was an apartment building or an individual home. We do know that Bouchet's residence was in the middle of the predominately Afro-American wards of St. Louis, only ten blocks from Sumner High School, and directly across the street from Provident Hospital.[42] The significance of Bouchet's address, however, lies in the window it provides into his life and activities while living in St. Louis.

Bouchet lived in an Afro-American neighborhood referred to as Mill Creek, a fifteen square block area just west of Union Station. Afro-Americans began moving into the area during the 1850s and by 1900 more than half of the 35,000 Afro-Americans living in St. Louis resided in the Central City and Mill Creek areas.[43] The neighborhood quickly became the center of St. Louis Afro-American culture. Josephine Baker was born there and the musical traditions of ragtime and the "St. Louis Blues" emerged in the Mill Creek area through popular local musicians, Scott Joplin and W. C. Handy.[44] One could speculate on whether Bouchet spent his evenings listening to Scott Joplin performing at the Rosebud Bar, the "Headquarters for Colored Professionals," located just six blocks from his residence.[45] More likely he took in Joplin's opera "A Guest of Honor" that had two performances at the True Reformers Hall in the summer of 1904.[46]

Friendships and associations are important indicators of a persons values, attitudes and goals. And, evidence clearly indicates that Edward Bouchet was associated with some of the most influential Afro-Americans in St. Louis at the time. Most notable of Bouchet's associates was Walter M. Farmer, a highly respected and accomplished lawyer. Born in Brunswick, Missouri in 1867, Farmer attended Lincoln University and was the first Afro-American to enter and graduate (cum laude) from the Washington University Law Department in 1889. As a practicing attorney, Farmer was also the first Afro-American attorney to argue a case before the Missouri State Supreme Court, and eventually presented the case before the United States Supreme Court. A Republican, Farmer was a delegate to the National Republican Conventions of 1896 and 1904, as well as being a member of the NAACP.[47]

How and in what capacity Bouchet came to be acquainted with Farmer is unknown. What is known is that at least some of the time during Bouchet's stay in St. Louis, he and Farmer shared a residence as the 1904 St. Louis City Directory lists both Farmer and Bouchet living at 2724 Morgan Street.[48] This fact alone provides some basis for speculation on the nature of their relationship. Bouchet was probably introduced to Farmer by Charles F. Joy, a college classmate of Bouchet's at Yale. By the time Bouchet made his way to St. Louis, Charles Joy had become a prominent trial attorney in St. Louis and had just finished serving his fifth term in the United States Congress.[49] As a prominent St. Louis lawyer and political leader, Joy would have known Farmer and been aware of his status within the Afro-American community. In all likelihood, Bouchet contacted Joy who then introduced the two men upon Bouchet's arrival in St. Louis. And, recognizing that Bouchet was in need of accommodations, Farmer probably offered him a room at his residence. However the connection between Farmer and Bouchet came about, it put Bouchet in contact with many other prominent Afro-Americans in St. Louis.

One surviving piece of evidence in support of Bouchet's involvement with St. Louis's Afro-American elite pertains to the issue of Jim Crow segregation at the 1904 World's Fair. From the very beginning of the Exposition and continuing throughout the spring and early summer of 1904, incidents of discrimination against Afro-Americans at the Fair were being reported in the local and national press. Most of the complaints pertained to specific events in which Afro-Americans had been denied admittance into a restaurant or refused service at the various concession stands.[50] The Louisiana Purchase Exposition was being billed as a distinctly universal Exposition, were all people would be on an equal footing, with no discrimination based on race or sex.[51] The Exposition company publicly proclaimed that the grounds of the Exposition would be open to all and the facilities under their direct control would be free from any restrictions.[52] However, many of the food and drink concessions were run by private companies who were free to set their own policies. As more and more incidents of overt discrimination reached the press, the Afro-American community in St. Louis, primarily through the Forum Club, began to stage protests against the Jim Crow regulations operating at the Exposition. The Forum Club even brought W. E. B. Du Bois to give a public speech speech in which he denounced the racism existing at the World's Fair.[53] The Exposition company immediately found itself facing a serious public relations disaster.

In an effort to head off any major demonstrations at the Exposition, the Executive Committee of the Exposition company, created an Afro-American Bureau. The bureau would be housed in a small building that would provide refreshments and bathrooms facilities, a place to rest, and would be staffed by an Afro-American hostess.[54] In addition to setting up the Afro-American bureau, the Committee on Ceremonies for the Exposition finally agreed to earlier demands from the Afro-American community that a day be set aside to be known as "Negro Day" or "Emancipation Day." The Exposition company authorized the formation of a committee to organize such a day and gave them August 1, 1904 as "Negro Day" at the World's Fair.[55]

The Committee on Negro Day included seventeen of the most prominent Afro-American men in St. Louis.[56] The Chairman was prominent attorney Walter Farmer and the committee was headquartered at his law office at 1107 Clark Avenue. Vice-Chairman was Reverened R. E. Gillum, pastor of the Union Memorial United Methodist Church, the oldest Methodist Church for Afro-Americans west of the Mississippi. David E. Gordon was the committee's secretary and, at the time, Principal of the L'Ouventure School. And, Edward Bouchet was the committee's Treasurer and also served as Chairman of the Programme sub-committee. Another member of the committee was William H. Huffman, a colleague and fellow teacher with Bouchet at Sumner High School. Huffman later went on to become a high school principal and eventually became the first Vice-President of Stowe Teachers College in St. Louis. Also on the committee was Dr. Thomas A. Curtis, one of the first Dentists in St. Louis. Dr. Curtis served on many local race relations groups including he St. Louis Community Council, and was a former president of the NAACP. Other members included Edward Williams, Principal at the Dessalines School; Arthur Freeman, Principal at the Wheatley School; John B. Vashon, Principal at Attucks School; Richard H. Cole, Principal at Simmons School; and Physicians, Dr. William H. Mansifee and Dr. Ottoway T. Fields.[57]

For Bouchet, St. Louis had much to offer. He was teaching science at a high quality high school that stressed academics over vocational education. The Afro-American neighborhoods were vibrant and organized, offering a vast array of cultural and social activities. He was actively involved in community affairs and his friendships brought him into constant contact with the best and most successful members of the St. Louis Afro-American community. Yet, he must of have felt out of place, something must have been missing from his experience in St. Louis. For after less than four years, Edward Bouchet got on a train and made the 1000 mile trip back to Connecticut. The reasons he left St. Louis and returned home to New Haven in March of 1905 are and will probably remain a mystery.

In October of 1906 Bouchet resumed academic work as he secured a teaching and administrative position at St. Paul's Normal and Industrial School in Lawrenceville, Virginia.[58] St. Paul's was founded by James S. Russell in 1888 as an elementary and secondary school affiliated with the Episcopal church. At the secondary level, the school primarily offered a teacher education program with algebra, geometry, English, geography, history and music the main courses of instruction. Bouchet was hired as the Director of the Academic Department, but also taught physics, chemistry, Latin, and civics. St. Paul's apparently had escaped the movement toward vocational education as the school's several hundred acres of farm land and extensive industrial equipment were rarely used for educational purposes.[59]

Information supplied by two residents of Lawrenceville offers a personal recollection of Bouchet while he was at St. Paul's.

> "He was a very handsome man, highly intelligent and proud. He was envied by the local citizenry (white) because of his intellect, and because he was not an 'Uncle Tom,' but he was very well liked and respected by the students and faculty at St. Paul's"[60]

Nevertheless, Bouchet was Afro-American and his intellect and academic status could not prevent him from being a victim of racism. As recalled by the above informants, Bouchet was assaulted by one of the towns prominent lawyers as a result of being accidentally bumped by Bouchet in going around a corner of one of the city streets.[61] Whether the assault was a factor is open to speculation, but Edward Bouchet left St. Paul's in June of 1908 and once again returned to his family home in New Haven.

In August he was accepted for the position of principal at Lincoln High School in Gallipolis, Ohio. He served in this capacity for five years, leaving in 1913.[62] Undocumented information indicates that after leaving Ohio, Bouchet may have taken a position at Bishop College in Marshall Texas, but was forced to retire from teaching in 1916 because of ill health. He returned to New Haven for the last time as he died at the age of 66 in his boyhood home at 94 Bradley Street on October 28, 1918.[63]

The Promise Denied

The obvious question that comes to mind in looking at the life of Edward Bouchet is why? Why was this gifted and highly educated scientist denied the opportunity to utilize his qualifications in the pursuit of scientific study and research? Why was he denied a position at one of the growing graduate programs that could have used someone of his ability, education and training? Why did he spend an entire lifetime teaching in high schools where his talents were underutilized? The obvious answer, of course, is racism. Edward Bouchet had the misfortune to be Afro-American during an era in which American was divided into two separate and unequal parts; one white where the promise of education could be realized, the other Afro-American where the promise was denied by law and violence. Regardless of this brilliance or his educational achievements, Bouchet was still Afro-American. To place this into proper perspective, it should be recalled that from 1885 to 1915, 3500 Afro-Americans were lynched; 235 in 1893 alone.[64] Also, beginning in 1875, Jim Crow segregation laws appeared in Tennessee and quickly spread across the south and into the north as well. This was the wrong time to a talented and educated Afro-American in a color-based caste society. But, this all too obvious answer begs the question as other factors played an important role in denying Bouchet the promise he so rightly deserved. Paradoxically, the other influence that had such an impact on the life of Edward Bouchet came not from white society, but from another Afro-American man, Booker T. Washington.

In 1872, while Bouchet was halfway through his undergraduate program at Yale, a sixteen year old Booker T. Washington arrived at

Hampton Institute. Founded by Samuel Armstrong, Hampton embodied the ideas of practical education and taught its students that labor was a "spiritual force" and emphasized the need to acquire vocational skills.[65] Strongly influenced by his experience at Hampton, Booker T. Washington took the concept of vocational education with him and founded Tuskegee Institute were women were taught cooking and sewing, and the men learned the trades of carpenters, blacksmiths and plumbers.[66] The Washington doctrine of industrial education was hailed by whites in the north and south. White society liked Washington's conciliatory approach and relative disinterest in political and civil rights for Afro-Americans. More importantly, they agreed with his advocacy of a form of education that they believed would keep Afro-Americans in their proper place in society.[67]

Openly hostile to academic education for Afro-Americans, Booker T. Washington traveled across the country promoting the virtues and successes of the Hampton-Tuskegee model of vocational education. Although not the originator of this form of education, he was its foremost proponent. Washington's prestige grew to the point where he was viewed by whites and many Afro-Americans as the leading spokesperson for the Afro-American community. Anytime political or educational leaders needed counsel on a question of race and education, Washington was called upon to provide the 'proper' advice. So influential did he become, that before white philanthropists would make donations to Afro-American schools, they sought Washington's assurance that their money would be spent for industrial education.[68] Given the poor financial condition of most Afro-American schools at the time, it is not surprising that many succumbed to the Washington philosophy and moved to develop vocational curricula in return for financial support.[69] And, it was this rising tide of enthusiasm for vocational education that would encircle Bouchet and prevent him from achieving anything more than a footnote in the history of Afro-American education.

Edward Bouchet spent twenty-six years at the ICY before he embarked on his nomadic search for a meaningful academic position. Bouchet obviously found something in Philadelphia that made him spend the better part of his life there. What he found was an academic atmosphere that gave support to the educational philosophy that had become part of his consciousness since his days at Hopkins Grammar School. He had been recruited to the ICY by Alfred Cope, a strong advocate and believer in the worth of academic education for Afro-Americans. Cope hired and promoted to principal Fanny Jackson Coppin who also supported the academic approach to Afro-American education. Under this leadership, Bouchet could not have been anything but optimistic as to the future of the ICY and of his role in its growth. However, that initial optimism surely faded during the late 1890's as the specter of industrial education appeared on the horizon and in the future of the ICY.

Booker T. Washington made many visits to Philadelphia and discussed his ideas of vocational education in public forums. The managers of the ICY were impressed with what he had to say and began to introduce changes in the direction of the school. Jackson tried to fight these moves, but the managers were in complete control of decisions concerning school policy. In the face of this uphill battle, Jackson decided in 1902 to resign from her position at the school. In locating a replacement, the Board of Managers consulted with Booker T. Washington and Hollis B. Frissell, the principal of Hampton Institute. Upon the recommendations of these two leaders of vocational education, Hugh Browne was appointed the new principal. To facilitate the change to the vocational model, all the teachers were fired and plans were made to move the school to a rural location outside the city.[70]

That Edward Bouchet was directly involved in this controversy can be detected in the minutes of the manager's meetings in which a report by the new principal, Hugh Brown, concerning the "Bouchet problem" was never entered into the written record.[71] Whether it was a protest over the change itself, the fact that he was not considered for the vacant position of principal, or some combination is unknown. Regardless, the board saw fit to punish Bouchet when they authorized three months severance pay to all the fired teachers except Bouchet.[72]

The extent of Bouchet's position in the debate between academic and vocational education is not known. Yet, it is instructive that the positions he secured after leaving the ICY were in schools that maintained the prominence of the academic tradition with which he was so familiar. Whether this was by choice or necessity is open to question. Given his educational philosophy, he may have been unwelcome at those schools that had adopted the vocational model. Insight into this matter can be found in two letters Bouchet wrote to Hollis Frissell of Hampton Institute inquiring about the possibility of joining the faculty. One could speculate that if Bouchet had strong adverse feelings about industrial education, he would not have made the effort to secure employment at the very school that initiated the vocational education movement. Unless, desperate for employment, pragmatism overcame personal philosophy.

Frissell wrote back to Bouchet indicating that he would like to have him on the faculty at Hampton, but a position was not available. Frissell also mentioned he would write Booker T. Washington concerning Bouchet's search for a teaching position, adding he would be glad to help in any way towards finding the "right place" for him as he was greatly interested in his success.[73] In a subsequent letter to Washington, Frissell inquired if their might be a place for Bouchet at Tuskegee, again indicating he would have liked to hire him at Hampton if only a position had been available.[74] Bouchet never taught at Tuskegee, although there is no evidence to indicate if a position had ever been offered.

There is a special irony in the relative position of Frissell vis a vis Bouchet: they had been classmates at Yale. After graduating from Yale, Frissell attended divinity school and eventually became chaplain at Hampton Institute in 1880. When General Armstrong, Hampton's founder died in 1893, Frissell became principal and managed the school until 1917. One can only wonder what impact such a turn of events must have had on Bouchet. If anyone was qualified to be the head of a major Afro-American educational institution it was Bouchet. But given the racial climate of the 1890's the position went to a white college classmate with average academic credentials.

With the ascendancy of the Hampton-Tuskegee vocational model as the dominant social philosophy for educating Afro-Americans, Bouchet found that he was now unwanted at both Afro-American and white colleges. Given the prevailing racism of the time, no white college would have seriously considered him for a position on their faculties even with his superior academic qualifications. At the same time, his classical orientation for academic education and skills in the natural sciences made him increasingly unattractive as a candidate at Afro-American colleges that had adopted the vocational curriculum.

What Might Have Been

One can only surmise Edward Bouchet's career path had he been white. Given his Yale education, superior intellectual ability, and one of only a handful of Ph.D. physicists in the country, it is clear he would have been one of the leaders in the development of scientific research and teaching in America. A brief look at a few of Bouchet's classmates at Yale supports this argument.

Charles Benton went on to earn a Litt.D. degree from the University of Pennsylvania and was Professor of French Literature at the University of Minnesota from 1880 until 1913. Henry Farnam went on to earn a masters degree and a LL.D and would become president of the American Economic Association from 1910 to 1911. William Halsted received a MD. degree from Columbia Medical School and became a professor of surgery at Johns Hopkins University. Edward Morris earned a Litt.D. degree from Harvard and was a professor at Williams College from 1885 until 1891. In 1916, he served as the president of the American Philosophical Association. And, Edgar Reading received both an MD. and a LL.D degree and served as President of Bennet Medical College from 1912 to 1914.[75] Edward Bouchet ranked above all these individuals academically at Yale. Yet, at the age of fifty, while many of his white classmates were at the top of their respective professions, Bouchet was an overqualified high school teacher.

The extent to which these turn of events affected Bouchet personally is unknown. He may never have desired to teach at a major university or work at an emerging research facility. Rather, his goal may have been to continue what Sarah Wilson had begun in New Haven; to provide the best education possible to his fellow African-Americans. But, the impact of Booker T. Washington must have been a disappointment for Bouchet. The vocational education movement sent Afro-American students on a detour from which they have yet to recover. Had teachers such as Edward Bouchet and Fanny Jackson Coppin been the leaders in maintaining the direction of Afro-American education into the 20th Century would we be grappling today with the gap between white and minority students in our schools? If academic education and schools such as the Institute for Colored Youth not been derailed by vocational training, would we today be facing the problem of underrepresentation of Afro-Americans and other minorities in science and engineering?

Edward Bouchet was first. The first Afro-American Phi Beta Kappa, the first Afro-American to attend and graduate from Yale, the first Afro-American to earn a Ph.D. from an American University, yet he was last in line for the rewards society usually bestows on those few who perform at such a superior level. Edward Bouchet was not a social activist like W. E. B. Du Bois; was not a creative research scientist like Charles Drew or Percy Julian; was not a published scholar like Ernest Just or Charles Turner, but Edward Bouchet deserves to be more than a footnote in the history of Afro-American education. His name belongs among the pioneers of Afro-American science and education for the part he played as a role model and inspiration to hundreds of students who sat in his classroom or worked in his laboratory.

Outside the walls of the ICY Afro-American youths were being told they were unable to appreciate or incapable of learning anything but the most rudimentary vocational skills. But, in Dr. Bouchet's classroom, they had before them a living contradiction to racist claims of innate Afro-American inferiority. It is doubtful that the full impact Edward Bouchet had on Afro-American education will ever be know; that he had an impact is undeniable. Mrs. Lillian Mitchell Allen, former chairwomen of the Department of Music Education at Howard University, remembers Dr. Bouchet from her childhood days in Gallipolis, Ohio. She states that as perhaps the most highly educated person in the area, he inspired both Afro-American and white young people with hitherto unknown goals. Mrs. Allen recalls that her brother, Dr. J. Arnot Mitchell, was influenced by Bouchet. Arnot Mitchell went on to study at Bowdoin College (Phi Beta Kappa, 1912) and eventually became the first Afro-American faculty member at Ohio State University.[76]

From his own personal accomplishments and those of his students, the absurdity of the claims being made by the proponents of vocational education concerning the supposed inability of Afro-Americans to appreciate an academic education could not have been more obvious to Bouchet. It probably never occurred to him that Afro-Americans could not master the fields of classical education and science. Even in the face of opposition and a changing public mood on Afro-American education, Edward Bouchet never altered his personal or educational ideals.

Epilogue

Edward Alexander Bouchet, age 66, died at his mother's home on October 28, 1918 from complications associated with high blood pressure. His funeral was held at St. Lukes Protestant Episcopal Church, with burial in the family plot (Path H, #10) in Evergreen Cemetery, New Haven Connecticut. At the time of his death he was survived by his mother, Susan Cooley Bouchet, two sisters, Francis (Fanny) Turner and Georgianna Bouchet, and two nephews, Thomas B. Jones and Otto C. Turner. A third sister, Jane Jones, preceded Edward in death at the age of 57 on January 24, 1900. Susan Bouchet died on February 11, 1920 at the age of 102. Georgianna Bouchet died on October 23, 1924 at the age of 70. Francis Turner died on November 30, 1934 at the age of 89. Otto C. Turner, Edwards's nephew by his sister Francis, was born on August 9, 1873 and died on March 20, 1928 at the age of 55. Otto had married a Clara E. Downing, born at Pittsfield, Massachusetts, and together they had a son, Lewis Turner, born on May 18, 1903. Thomas B. Jones, Edwards's nephew by his sister Jane, was born on July 30, 1872 and died in 1929. Edward Bouchet never married, so any living relatives of Edward Bouchet would have to be descendents of his nephews Thomas Jones, Otto Turner and Lewis Turner.

Forgotten for too long, it is hoped that this book will shed light on Edward Bouchet; who he was as a man, a teacher, and mentor and role model to thousands of young Afro-American men and women. For a man who gave forty years of his life to serving the Afro-American community, it is unfortunate that so little of the richness and detail of his life and work remain. It is hoped that other historians of Afro-American culture, professional and amateur alike, will pursue the effort to fill in the gaps that remain and uncover more about the life of Edward Bouchet. We would so much like to know more about the man who Mrs. Sarah Isaac, (who in 1976 was the oldest living graduate of the Institute of Colored Youth and died in 1980 at the age of 100), remembered as a handsome, well-dressed, and well-liked professor; and for whom just the mention of his name brought back pleasant memories and a smile to her face.[77]

1. J. H. Benham, Benham's New Haven Directory, no. 34 (New Haven: J. H. Benham, 1873; U.S. Census, New Haven Connecticut); 7th Ward, p. 46.

2. Benham, Benham's, New Haven Directory, 1895–1920.

3. New Haven Colony Historical Society, "Proceedings of the Board of Education from July 11, 1861 to May 31, 1866," vol. 11, MSS B-26, Box 5, Folder C.

4. Annual Catalogue of the Officers and Scholars of the New Haven Hopkins Grammar School at New Haven, Connecticut, July 1870. (New Haven: Tuttle, Morehouse & Taylor, 1870), p. 9.

5. Ibid., p. 20.

6. Yale College Student Grade Books, "Grades: Class of 1874," vols. 4 & 5, Yale University Manuscripts and Archives, New Haven Conntecticut.

7. Statistics of the Class of 1874 in Yale College. (New Haven: Tuttle, Morehouse & Taylor, 1874).

8. Linda Marie Perkins, "Fanny Jackson Coppin and the Institute for Colored Youth: A Model of Nineteenth Century Black Female Edcuational and Community Leadership" (Ph.D. dissertation, University of Illinois at Urbana-Champaign, 1978), p. 116.

9. Nancy N. Soper, et.al., Physics Doctorates of Yale (New Haven: Yale University Press, 1976), p. 4.

10. Ibid.

11. Perkins, "Fanny Jackson Coppin," pp. 60–64.

12. Charline Ray Howard Conyers, "A History of the Cheyney State Teachers College, 1837–1951" (Ed.D. dissertation, New

York University, 1960), p. 151.

13. Perkins, "Fanny Jackson Coppin," p. 79.

14. Ibid., pp. 115–116.

15. Ibid., pp. 57–58.

16. Institute for Colored Youth, "Minutes of the Board of Managers," vol. RS 517, Richard Humphreys Foundation, Friends Historical Library, Swarthmore College, Swarthmore, Pennsylvania.

17. Philadelphia City Directory.

18. Conyers, "Cheyney State," pp. 162 & 177.

19. Institute for Colored Youth, "Managers Minutes," Box 5 & 6.

20. Ibid.

21. Ibid.

22. Ibid.

23. Ibid.

24. Yale Obituary Record, Part Fifth, 1909–1919 (New Haven: Price, Lee & Adkins, 1919), pp. 82–83.

25. Roger Lane, William Dorsey's Philadelphia and Ours: On the Past and Future of the Black City in America. (New York: Oxford University Press, 1991), p. 234.

26. Yale Obituary Record. Part Fifth, 1909–1919, pp. 82–83.

27. Lane, William Dorsey's Philadelphia and Ours, p. 279.

28. Ibid., p. 281.

29. Ibid., p. 289.

30. Ibid.

31. Ibid., p. 290.

32. Perkins, "Fanny Jackson Coppin," p. 211.

33. Ibid., p. 216.

34. Ibid., p. 224–225.

35. Ibid., p. 244.

36. Sumner Centennial Steering Committee, History of the Charles Sumner High School: Centennial Edition (St. Louis: Saint Louis Public Schools, 1975), pp. 11–12.

37. Ibid., p. 13.

38. St. Louis Globe-Democrat, Tuesday, 20 February 1979, p. 14D.

39. Obituary Record of Yale Graduates, 1918–1919, p. 19–20.
40. Ibid.
41. Ibid.
42. Gould's St. Louis Directory (1903, 1904, 1905), Missouri Historical Society, St. Louis Missouri.
43. Katharine T. Corbett & Mary E. Seematter, "Black St. Louis at the Turn of the Century" Gateway Heritage 7, no. 1 (Summer, 1986), p. 41.
44. Issac D. Darden, "African-American Communities and Institutions" Proud 12, no. 1 (1981), p. 12.
45. Corbett and Seematter, "Black St. Louis," p. 47.
46. Nathan B. Young, "Nineteen Years to Go" Proud 12, no. 1 (1981), p. 26.
47. Joseph J. Bris (ed.), Who's Who in Colored America vol. 1927, p. 64–65 (Missouri Historical Society, St. Louis Missouri).
48. Gould's Directory (1904).
49. Walter B. Stevens, History of St. Louis: The Fourth City, 1764–1909 Vol. II (Chicago: The S. J. Clarke Publishing Company, 1909), p. 958.
50. Louis Globe-Democrat, 2 June 1904, p. 3; 14 July 1904, p. 14.
51. H. S. Ruehmkorf, The Lousiana Purchase Exposition (n.d.), Missouri Historical Society.
52. St. Louis Globe-Democrat, 2 June 1904, p. 3.
53. Young, "Nineteen Years to Go," p. 25.
54. Louisiana Purchase Exposition Company Collection, "Executive Committee Minutes, 31 May 1904 to 1 July 1904" pp. 2608–2803 (Missouri Historical Society, St. Louis, Missouri); St. Louis Globe-Democrat, 2 June 1904, p. 3.
55. Letter, "Walter M. Farmer to Booker T. Washington," 8 April 1904. Booker T. Washington Papers, Container 802, Louisiana Purchase Exposition Company Collection, Missouri Historical Society, St. Louis, Missouri.
56. Letter, "Walter M. Farmer to Booker T. Washington," 1 August 1904. Booker T. Washington Papers, Container 288, Louisiana Purchase Exposition Company Collection, Mis-

souri Historical Society, St. Louis, Missouri.

57. Joseph J. Bris (ed.), Who's Who in Colored America vol. 1927; St. Louis Necrologies, vols. 16 & 24; Negro Scrapbook, vol. I (Missouri Historical Society, St. Louis, Missouri); UMSL Black History Collection, 1895–1983 (University of Missouri at St. Louis, St. Louis, Missouri).

58. Obituary Record of Yale Graduates, 1918–1919, p. 19–20.

59. St. Paul's College, The St. Paul Bulletin, vol. 1, no. 1 (Lawrenceville, VA: St. Paul Normal & Industrial School, 1907–1908).

60. Folder, "Edward A. Bouchet," H. B. Frissell Collection, Hampton University Archives, Hampton University, Hampton, Virginia.

61. Ibid.

62. Obituary Record of Yale Graduates, 1918–1919, pp. 19–20.

63. Ibid.

64. Benjamin Brawley, A Social History of the American Negro (New York: The Macmillian Company, 1970), p. 295.

65. John Hope Franklin, From Slavery to Freedom (New York: Alfred A. Knopf, 1973), p. 284.

66. Benjamin Quarles, The Negro in the Making of Ameica (New York: The Macmillian Company, 1969), p. 166.

67. Franklin, From Slavery to Freedom, p. 286.

68. Quarles, Making of America, p. 166.

69. Ibid.

70. Perkins, "Fanny Jackson Coppin," p. 285–286; Conyers, "Cheney State," p. 112.

71. Board of Mangers, Minutes?

72. Ibid.

73. Letter, "Frissell to Bouchet," 8 April 1902. H. B. Frissell Collection.

74. Letter, "Frissell to Booker T. Washington," 8 April 1902. H. B. Frissell Collection.

75. Catalogue of Graduates, vol. 2, 1701–1954 (New Haven: Yale University), pp. 217–218.

76. Folder, "Edward A. Bouchet" H. B. Frissell Collection.

77. Letter, "Linda Perkins to H. Kenneth Bechtel," 1 August 1987, personal correspondence.

Hopkins Grammar School (1860)
New Haven, CT

Ed. A. Bouchet.

Edward A. Bouchet, ca. 1869. This may be a Hopkins Grammer School
graduation photograph. Reprinted courtesy of Yale University Library,
Manuscripts and Archives.

Edward A. Bouchet at Yale University, ca. 1874. Reprinted courtesy of Yale University Library, Manuscripts and Archives.

Edward A. Bouchet at Philadelphia, PA (?), Yale University alumnus meeting, ca. 1902. Reprinted courtesy of Yale University Library, Manuscripts and Archives.

Edward A. Bouchet, ca. 1902. Reprinted courtesy of Yale University Library, Manuscripts and Archives.

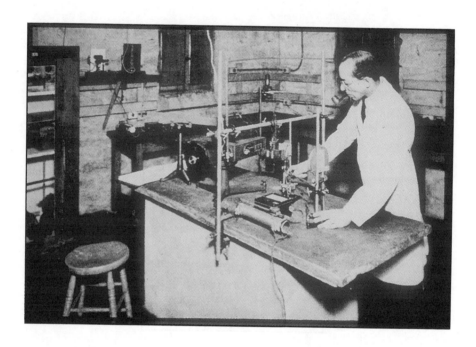

Elmer S. Imes in his physics laboratory at Fisk University (Nashville, TN), ca. 1935. Imes was the second African American to receive the doctorate degree in physics (University of Michigan, Ann Arbor, MI; 1918). Reprinted courtesy of Special Collections, Fisk University Library.

Willie Hobbs Moore (front row, right) was the first African American woman to earn the physics doctorate degree (University of Michigan, 1972). This photograph was taken with her family, ca. 1972. Reprinted courtesy of Sidney Moore.

The First Edward Bouchet International Conference on
Physic and Technology.

I.C.T.P.   Trieste

                Jun 9–11 1988.

Participants in the First Edward Bouchet International Conference on Physics and Technology. This was held at the International Center for Theoretical Physics, Trieste, Italy, 9–11 June 1988. Reprinted courtesy of ICTP, Trieste, Italy.

Chapter 4

In Search of Edward Bouchet*

John A. Wilkinson[†]

How many of you know the name of Edward Bouchet and his accomplishments? Virtually every black scientist or mathematician or graduate of Fisk, Tuskegee, Howard or similar institutions would tell you that he was the first black American to earn the Ph.D., at Yale, in 1876, in physics.

When I first thought of speaking to you of Bouchet, I was confident that it would be a relatively easy task to gather information about him, but I have rediscovered the reason students of history relish the detective novel. My paper tonight, consequently, is not unrelated to previous ones on Dick Francis, Emma Lathen, and the origins of the Hopkins Grammar School. It is an account of detection and has been retitled, "In Search of Edward Bouchet."

The basic facts of Bouchet's life are mostly known largely through the historical records of the Yale College Class of 1874; who Bouchet was, what he thought and said, how he influenced his students and colleagues, friends and associates has become, for me at least, a mystery.

I first heard of Edward Alexander Bouchet in the summer of 1974, as the Headmaster of Hopkins Grammar School. While going through records of the School from the 19th Century, I discovered

*Presented before the Social Sciences Club of New Haven, CT ca 1988.

†American University of Beirut, 850 Third Avenue, 18th Floor, New York, NY 10022–6297.

class photographs with an occasional black student. Imagine my surprise when I discovered that one of them, Bouchet, had graduated first in his Hopkins class of 1870, had gone on to Yale, where he was elected to Phi Beta Kappa in junior year, graduated 6th in his class, and two years later earned a Ph.D. in physics.

This seemed to be a man worth coming to know, and, over the last ten years, I have haphazardly made several efforts to become acquainted with him. I have learned much, and I have learned little. Let me explain.

Bouchet was born in New Haven on September 15, 1852, the son of William Francis and Susan Cooley Bouchet. The father, William, was, by one account, born in New Haven in 1817; in another, came to New Haven in 1824, from Charlestown, South Carolina, as the body servant of the father of Judge A. Heaton Robertson. The Judge was Edward Bouchet's contemporary in Yale College. (And I well remember A. Heaton Robertson, 3d, in the Yale College Class of 1965.)

William Bouchet was described as a man of exceptional character and force of personality who was an early pillar of the Temple Street Church, now the Dixwell Congregational Church.

The mother, Susan Cooley Bouchet, was born in Westport, Connecticut, on October 1, 1817, the daughter of Asher and Jane Drake Cooley. She died on February 11, 1920, the 103rd year of her life. She outlived her husband by 35 years.

Edward Bouchet was the only son and youngest of 4 surviving children. The family lived at 42 Bradley Street and later at 94 Bradley Street, and, even later, apparently at 42. Both houses were on the block of Bradley destroyed for the Trumbull Street connector to Interstate 91.

Lower Bradley Street, off State Street, then known as Negro Lane, was in what was regarded as one of the better Negro settlements in New Haven, for it housed those blacks who worked as servants for the affluent Yankee families of Hillhouse and Whitney Avenues. Though poor soil made successful gardening difficult, the Negro section along State Street was better than that of the Hill, known as Sodom Hill, with its partly drained salt marsh and the stench and noise of the tanyards along West Creek. It was also much superior to the area

Between Poverty Square (between Whalley and Goffe) and the Farmington Canal, and now known to us as the Dixwell and Newhallville neighborhoods.

A death rate of 38 per 1,000 for blacks in 1860 against that of 17 for whites tells much of the story.

These "People of Color" in New Haven were a recognized social caste, until and even after the Civil War. The New Haven directories until 1865 classified them with the word "col'd" after their names. They were assigned to special sections in churches, and attended separate public schools. "Branded with ignominy," in the words of Leonard Bacon, these free Negroes, numbering about 1,500 in 1860, were generally excluded from apprenticeships in the trades. Nine out of 10 black adults were a menial or laborer, and the competition for these positions was fierce with the new Irish immigrants.

By 1850, there were four primary schools for black children in New Haven, two near Poverty Square, one in Mt. Pleasant, adjacent to the Hill, and the oldest, founded in 1811, the Artisan Street Colored School, on the corner of Artisan and Court Streets, between State and Olive. It was there Bouchet went to school before entering Hopkins.

How he got to Hopkins I have not learned. Who paid (perhaps the Robertson family?), how he was treated, whether he had friends, these facts are not recorded. We do know that he graduated first in his class and delivered the valedictory address at graduation.

Twenty-four years later, when Bouchet was 42, he wrote to his Hopkins and Yale classmate, George L. Fox, then the Rector of Hopkins, from Philadelphia:

834 Lobard St.
Phila., Pa. 6-27-1894

My dear Fox: -

In accordance with your request, I send the following data:
Residence while at H.G.S.
94 Bradley St., New Haven, Conn.
Degrees obtained, B.A. '74
and Ph.D., '76, Yale University
Present occupation. Instructor in Chemistry and Physics,
Institute for Colored Youth,

Phil., Pa.

    With thanks for courtesies received,
        I remain
            Very truly yours,
                Edward A. Bouchet
                    H. G. S. '70

This, apparently, is the only surviving letter written by Bouchet. It is hardly the kind of letter one would expect, by today's customs, between men who had been classmates in two schools. Remember there were only 125 in the Yale College Class of 8174. And Hopkins probably had not graduated more than 25. But, then, 100 years ago formality was in much greater supply.

We know more about Bouchet's life in Yale College. Though, apparently, not the first black to enter Yale, he was the first to graduate.

In his freshman year, he had a stunning 3.36 average, with the highest grade of 3.52 in math. There was little variation in his performance during the next three years, though he fell below 3 with a 2.99 in German. In addition to his courses in the sciences (math, astronomy, mechanics, and physics) he also studied English, French, German, Greek, Latin, Logic and Rhetoric.

It was an interesting time for Yale in the post-Civil War period. Theodore Dwight Woolsey retired as President, after 25 years in office, and was succeeded by Noah Porter. Hazing of the Freshman Class ended. Young, that is, middle-aged, alumni were elected to the Corporation. Compulsory Chapel services were reduced from three per weekend to one. Farnam and Durfee Halls were built. A new Chapel, Battell, was started.

The class of 1874 was the second largest in Yale's history. The 125 students came from 18 states and 2 foreign countries. The largest group (28) was from Connecticut, 9 were born in New Haven, and the schools with the largest delegations were Hopkins 18, Andover 14, Hartford High School 7, and Exeter 3.

But what about Bouchet? We know that he lived at home. He was one of 4 students who never used a medical excuse during his 4 years as an undergraduate. He was one of only 2 classmates who never joined a secret society during any of the 4 years. The other,

Edward P. Morris, boasted of his independence. Bouchet and his classmates said nothing of Bouchet's independence or exclusion.

At Class Day and Commencement in June, 1874, despite his class rank of 6 and high orations or summa cum laude, he was not chosen to be one of the 16 speakers from his class. But neither was Kennedy, a New Haven Irishman who ranked 3rd, nor A. B. Thacher, son of a Yale professor, who ranked 5th.

At the Class Day Exercises of June 23, the Class sang several Yale songs and, also, "Three Little Darkies," and "Nellie (a prostitute) Was a (dark) Lady." It must not have been pleasant for Edward Bouchet.

The Commencement addresses by the graduates of 1874 retain their freshness. Robbins, the high stand man, spoke on "Goethe's Philosophy of Evil." Parkin, the salutatorian, addressed the Commencement in incomprehensible Latin, Curtis, #4, argued the "Justice of Edmund Spenser's Literary Reputation," and Whittemore, #8, gave a history of the "Taxation of Ecclesiastical Property." Charles Benton of Mt. Lebanon, Syria, spoke on the "War of 1860, between the Druzes and the Marianites of Mt. Lebanon." James Cadawalder Sellers, later to write Bouchet's obituary, exclaimed on the "Influence of William Penn Upon American Republicanism." The graduate photograph of that June shows Bouchet sitting at an angle different from his classmates, no one immediately beside him.

That fall, Bouchet entered the graduate program. While there he studied experimental physics with Wright, calculus with H. A. Newton, and, in the Sheffield Scientific School, chemistry and mineralogy with Allen and Brush. He completed his graduate work in only two years, and received the Ph.D. in 1876. Only 39 men had earned that degree between 1861, when the first in the nation was awarded by Yale, and 1876.

Bouchet's accomplishment was another first, but a specialist in "Measuring Refractive Indices" did not find a first job in a college of university. Instead, he became a teacher of chemistry and physics at the Institute for Colored Youth, a Quaker school, in Philadelphia, where he taught, without benefit of laboratory, for 26 years.

There is no evidence that Bouchet attended a Yale College reunion until 1909, 35 years after graduation. In 1889, his 15th reunion, he wrote for the class book that "There is every prospect that teaching will be my life-work." Ten years later, for his 25th reunion, he wrote, "Nothing worth especial mention has occurred in my life since 1889 ... I have endeavored to discharge my duty as a teacher to those coming under my care, and have aimed to be a good citizen, and to exemplify in my life the mottos of our Alma Mater."

In 1909, Bouchet came to his 35th reunion, and gave a much fuller account of his activities.

> "In September, 1876, I began teaching physics and chemistry in the Institute for Colored Youth, Philadelphia, Pa., and continued to fill that position until June, 1902. From September, 1902, until November, 1903, I was connected with the Sumner High School, St. Louis, Mo., as teacher of physics and mathematics. From November, 1903, until May, 1904, I was business manager for the Provident Hospital, a private institution located in St. Louis, Mo. From May, 1904, until March, 1905, I was United States Inspector of Customs at the Louisiana Purchase exposition in St. Louis, stationed at Ceylon Court. This appointment was obtained through the good offices of the Honorable Charles F. Joy and other St. Louis friends. In October, 1906, I became director of Academics at the St. Paul Normal and Industrial School, located at Lawrenceville, Va., where I remained until June, 1908, and in September, 1908, I accepted the position of principal of the Lincoln High School at Gallipolis, Ohio. My favorite recreations are walking and rowing. The classmates I have met most frequently have been George L. Dickerman, Henry W. Farnam, George L. Fox, George M. Gunn, Charles F. Joy, James C. Sellers and Edmund Zacher."

Why five jobs in six years after leaving Philadelphia he doesn't say. What his contacts were with Dickerman, Farnam (the Yale professor of 43 Hillhouse Avenue), Fox (the Hopkins Headmaster), and the rest of his classmates he doesn't say. Nor do they, except for James Sellers, the Philadelphian who spoke of William Penn at graduation, and wrote Bouchet's obituary.

Sellers gives us the only account of Bouchet's life in Philadelphia and suggests his reason for leaving the Institute for Colored Youth.

> "Bouchet spent twenty-six years of his active life in Philadelphia, where his memory is still cherished by many friends and

pupils. He taught in the Institute for Colored Youth, an old Quaker school, dating back to abolition times. About 1902 that school was merged into the Cheyney Training School for Teachers and removed to a rural location, about twenty miles from Philadelphia, where it is now conducted largely on industrial lines, after the models of Hampton and Tuskegee. Another of our classmates, Frissell, was much interested in the remodeled school and visited it on several occasions. But Bouchet was not in sympathy with many of the changes made and gave up his connection with the school to engage in other work."

You can imagine Bouchet, a student of the classics, pure mathematics, and experimental physics, not in accord with the educational philosophy of Booker T. Washington.

"While in Philadelphia Bouchet became connected with St. Thomas Church, one of the oldest colored Episcopal churches in the country, dating back nearly a century. He served on the vestry, of which body he was secretary, for a number of years, and was appointed by the bishop a lay reader in which capacity he frequently took part in the services of the church. He was greatly interested in the university extension work among the colored people of the city and did considerable lecturing himself at various times. Many of his pupils are now in active business and professional life, and all speak of him as a teacher with esteem and gratitude and bear affectionate witness to his high character and blameless life."

"During his residence in Philadelphia Bouchet took much interest in the local Yale Alumni Association and was a faithful attendant at its meetings and annual dinners. He won and retained the regard and kindly interest of its other members and was always received by them with cordiality and respect."

We also know from Sellers that Bouchet remained at Lincoln High School in Ohio for four or five years when a heart attack, or at least, what he calls an attack of the disease, arteriosclerosis, forced Bouchet to resign his position and return to New Haven for a prolonged rest. He returned to work as a professor at Bishop College, Marshall, Texas, but, in 1916, his health again forced him to return to New Haven where he died on October 28, 1918, at age 66.

Bouchet never married, but was survived by his mother, then 101 years old, two sisters and two nephews. His funeral service was at St. Luke's Protestant Episcopal Church, and he was buried in Evergreen Cemetery.

Sellers closes his obituary with great sensitivity:

> "Bouchet reflected great credit on his people and demonstrated by his own career that capacity to accomplish worthy things in intellectual fields. In all his associations, both in college and in later life, he showed himself the thorough gentleman. The memory of his quiet scholarly life will long remain as an influence for good among the members of his race and many others who were privileged to know him."

In the official Yale obituary, we learn that Bouchet was a member of the Franklin Institute and the American Academy of Political and Social Science. From his New Haven friend, the black lawyer and Corporation Counsel, George W. Crawford, we discover that Bouchet

> "was man of keen sensibilities, and unusual refinement. He was a prolific reader, and was greatly interested in the history of his own people and of his native town. He was a walking encyclopedia of New Haven people and events. He was a prolific collector of information about people and things in which he was interested, and his fine collection of scrap books evince his indefatigability as a collector of this sort of information. In his later years Dr. Bouchet took sides with the more progressive and radical leadership of his people. He was an enthusiastic member of the National Association for the Advancement of Colored People – a new, but influential organization, having for its object the most aggressive championship of negro citizenship rights."

As I come to the end of this account, I confess frustration, for I cannot say I know this man. Too much is missing. The Bradley Street houses are gone. The Ph.D. dissertation is missing from Sterling Library. His grave in Evergreen cemetery is unrecorded. His notebooks are lost.

Curtis Patton, Professor of Epidemiology at Yale, in the preface to a report of 1979 on the Status of Black Graduate & Professional Students, states:

> "Preeminent, humane, undaunted by enormous odds, untouched by self pity, Yale and a young man named Bouchet found their

resources and goals compatible more than a century ago. History was made. But no tradition was started."

"We may never know the specifics of Bouchet's suppression. We have no documents that give clues to his thoughts on his career. We know only that he lived during a period that can only be called terrible for Black people. His challenges must have been magnificent."

Chapter 5

African Americans Enter Science

Kenneth R. Manning*

Free blacks constituted the only group of blacks in ante-bellum America who had even a meager chance for exposure to educational opportunity and the scientific literature, and, by extension, possibly to engage in scientific work itself. It was through the efforts of the free black population that many black educational and religious institutions were able to grow and flourish in this period. Free blacks not only struggled for legal emancipation, they also were involved in liberation efforts, through the underground railroad, through marriage, and through purchase. In addition, following the passage of the Fugitive Slave Law in 1793, they all had to be concerned about maintaining their own personal freedom. For many, scientific work was a mere dream that they could only realize under more favorable circumstances and in a more congenial racial climate.

Indeed, free blacks who pursued higher education did so at New England colleges and tended to concentrate on the traditional disciplines of law, theology, and letters-disciplines that were of immediate practical use in the social and political agenda that consumed much of their attention. Almost no blacks took up science. Amherst College, Bowdoin College, and Dartmouth College led the way in accepting black students between 1820 and the Civil War. Dartmouth College had a special role as an institution of higher learning accessi-

*Massachusetts Institute of Technology, Program in Science, Technology, and Society, Room 16-236, Cambridge, MA 02139 (krmanning@mit.edu).

ble to people of color. None, however, could match the commitment that Oberlin College in Ohio maintained throughout the nineteenth century. A few black students pursued qualifications in medicine at places like Rush Medical College and the Harvard Medical School.

By starting newspapers and by setting up grade schools for colored youth, educated black individuals coming out of these institutions used the privileges they had earned, or had been awarded, to help build a small but important infrastructure for blacks. Free blacks were instrumental in developing their own institutions of higher education in the 1850s. Even so, the curricula developed at these colleges were primarily classical, the bulk of the course offerings being in the humanities and only a very few in the sciences. Thus, by the start of the Civil War, no blacks had engaged in pure science as such since the time of Benjamin Banneker at the end of the previous century. They were consistently active in the field of invention, however, attested to by the careers of Thomas L. Jennings, Henry Blair, George Peake, James Forten, Norbert Rillieux, and others who acquired patents during the early nineteenth century.

By the middle of the nineteenth century, American scientific institutions and organizations had begun to develop. White American scientists, working on a level of achievement and sophistication comparable to that of Banneker, were establishing impressive institutional structures on which their sons-and by now some of their daughters-could build and from which they could benefit. American science enjoyed the participation of educated individuals who were beginning to understand the works of Europeans in the vanguard of modern science. Rensselaer Polytechnic Institute was founded in 1824, The Lawrence Scientific School at Harvard in 1847 and the Sheffield Scientific School at Yale struggled under a different name during the 40s and was finally funded in 1861. The American Association for the Advancement of Science was chartered in 1848. In the 1850s, Harriet Beecher Stowe was writing her *Uncle Tom's Cabin*, hoping to contribute to the emancipation movement. Around the same time, Darwin was publishing his *Origin of Species* while Alexander Bache and others of the so-called Lazarroni were planning the National Academy of Sciences. In 1861, William Barton Rogers obtained a charter and a plot of land in Boston's Back Bay to develop "a

School of Industrial Science,"—the Massachusetts Institute of Technology. By the time of the Civil War science had definitely matured in the America, fully embedded in a social and political context, though opportunities for any kind of education for blacks, scientific or otherwise, were uncertain at the time.

Only with the Emancipation Proclamation of 1863 did any significant participation of blacks in the social, cultural, educational—and scientific—institutions of the country become at least possible. The Freedmen's Bureau, set up in 1865 to provide opportunities for former slaves and poor white men and women, was one of the first formal mechanisms by which blacks as a group benefited educationally and professionally. At the time, it was indeed the only way for most blacks to begin to participate minimally in scientific activity. Right after the end of the Civil War, black colleges and universities began to be set up, providing blacks with an essential start in their pursuit of education by awarding bachelor degrees. Such schools as Howard University, Atlanta University, Hampton Institute, and Fisk University, among others, were founded during this period and funded by the Freedmen's Bureau and religious organizations in the North-the Baptists, Methodists, Presbyterians, and Episcopalians. Private philanthropy through large educational foundations joined these activities to encourage the education for African Americans in all academic fields. The Peabody Education Fund and the John F. Slater Fund were two early ones that supported the black colleges and universities in this early period of their founding.

Established on 2 March 1867 Howard University set as its goal in its original charter "the education of youth in the liberal arts and sciences." Yet for a half century the sciences remained a relatively low priority in the curriculum. The largely white administration and faculty saw blacks as primarily teachers and preachers among blacks, with little opportunity to do research in science or even science teaching at a moderately advanced level. The first science courses were offered in the medical department, which opened in November 1868, a year and a half after the normal (teacher training) department. Although the Howard administration expressed a commitment to providing "suitable prominence to the several branches of physical science" in its Catalogue for 1873–74, the number of science courses

grew slowly. The educational opportunities for African Americans
at black colleges and universities were indeed limited at the early
period of their development.

In 1876, the first black Ph.D. from an American university, Edward Bouchet, received his degree in physics at Yale University. He
was one of the first recipients of any color to earn the Ph.D., the first
in America having been awarded just ten years earlier. Bouchet's
subsequent career did not, however, include research in the sciences;
instead, he became and remained a high-school teacher of science at
the Institute for Colored Youth, started by free blacks in Philadelphia in 1837. Professional opportunities in science were not open
to him, though he had worked beside some of America's top physicists, including the eminent Josiah Willard Gibbs at Yale. Similarly,
Alfred O. Coffin received a Ph.D. in biology from Illinois-Wesleyan
University in 1889, but went on to teach romance languages at a
black college. Bouchet's and Coffin's were nonetheless important accomplishments, counter-examples to the firmly held and widespread
belief in the mental inferiority of blacks.

It is the year 1876 that marked the end of Black Reconstruc-
tion in the South. The election of Rutherford B. Hayes that year
brought in the period of so-called White Redemption. Bouchet's career began during perhaps the most repressive period in American
history. For a quarter century, from 1876 to the end of the century,
marked the period of black disenfranchisement and the setting up
of jim crow laws. While the outlook for blacks in American society
was threatened from many vantage points, their educational efforts
were nonetheless strengthened as more black colleges and universities
emerged and continued to provide some opportunity for the development of a talented tenth, as characterized by W. E. B. Du Bois a
few decades later.

Since the early nineteenth century, free blacks had begun moving
into the medical profession in greater numbers as an outlet for their
scientific interests. Medicine offered a career in which the educational
requirements were not as extensive or as demanding as those necessary for a career in research science. Medicine fulfilled, in a direct
way, a notion of community service, as expressed in the educational
philosophy of Booker T. Washington. Moreover, a medical career for

a black was almost invariably carried out in the black community—
to clinically treat blacks, to help cure blacks, to protect whites from
diseases in the black community, and to avoid tainting the complexion of white professional institutions. Washington himself had said,
in an address delivered at a meeting in Atlanta in 1895, that the
races should be kept as separate as the fingers on a hand. At this
same meeting, blacks in medicine started a national organization
and began to develop a professional identity. The National Medical Association, founded in 1895, was the black counterpart of the
American Medical Association, which, effectively barred most blacks
from becoming members. The official political separation came in
the form of the historical Supreme Court case, Plessy v. Ferguson,
which declared separate but equal to be the law of the land. Educational institutions at all levels in the South fell into that framework.
So, black scientists who emerged in the early twentieth century were
products of a segregated educational system. By and large, they
would pursue their careers within that system.

It was not really until the turn of the twentieth century that a
handful of blacks began to enter the scientific fields. Among these
people come to mind Charles Henry Turner, George Washington
Carver, Ernest Everett Just, St. Elmo Brady, Elmer Imes, Julian
Lewis, and a little later, Percy Julian and Charles Drew. This co-
hort represents the first group of black scientists who received Ph.D.'s
from major white universities, pursued science at the research level,
and published in the leading scientific journals of the day. Prior to
the Second World War, their professional lives were almost exclusively tied to black colleges and universities.

One of the earliest opportunities that black scientists had to pursue research was in seasonal laboratories such as the Marine Biolog-
ical Laboratory at Woods Hole, which brought them directly into
a white environment. Among the blacks who worked there during
the summer season were Charles Henry Turner, E. E. Just, Samuel
Milton Nabrit, and, a female zoologist, Roger Arliner Young. Often
confronted by the prevailing racial attitudes of the time, these scientists nevertheless managed to engage in pioneering work at such
laboratories. E. E. Just accomplished first-rate results in embry-
ology, despite the subjection of himself and his family to a hostile

environment and racial slurs both in a scientific and non-scientific context.

Science was a deeply felt, personal commitment for Just, one that he finally left America to pursue in Europe. But his experience at Woods Hole, disturbing as it was in some ways, nevertheless paved the way for other blacks to go there. The doors of the community had been opened, and after the 1920s it was no longer as shocking to see a black researcher walking around the laboratory halls and out in the streets, or joining in the extracurricular activities and the social life of the place. One of the hardest things for blacks in these all-white communities was to locate housing or lodging.

During this same period, the first part of the twentieth century, Jews came into their own in science, first mainly in Germany and later in America. The associated institutes of the Kaiser-Wilhelm Gesellschaft in Berlin-Dahlem, founded in 1910, permitted scientists of Jewish heritage such as Richard Goldschmidt and the Nobel laureates Otto Meyerhof and Otto Warburg to engage in pioneering scientific research prior to World War II. Although anti-Semitism in the early twentieth-century scientific community in this country was prevalent, a quota system served to limit admissions to graduate schools and the hiring of faculty at major universities, but not to bar Jewish scientists from these places completely. With the reputations of Einstein, Michelson, and others to support them, Jewish students and scientists could be found spread out among major American universities and research institutions, and serving in some key scientific and administrative roles. Many carried on a tradition of acute social awareness and moral energy that had been hallmarks, for example, of Einstein and of the eighteenth-century black American scientist Banneker—a tradition generated in part by their personal experience of ethnic and racial intolerance. In this connection I think particularly of Jacques Loeb, the Flexner brothers—Abraham and Simon, Selig Hecht, and Robert Oppenheimer, to name a few. Thus, a critical mass insured the inclusion of other Jews.

It is indeed ironic that it was the Second World War that brought some public notice to black scientists. Before that time, they had worked individually and at black institutions, their number and presence not yet having really been strongly felt or observed in the scientific community. At Los Alamos and in the various branches of the Manhattan Project under way at the University of Chicago, Columbia University, and several research laboratories, some white scientists witnessed for the first time a sizable portion of black physicists and chemists entering their world, being mobilized as part of the scientific war effort. Blacks who worked on the bomb project included Moddie D. Taylor, Edwin R. Russell, George W. Reed, and the brothers William J. Knox and Lawrence H. Knox. In a talk at the American Physical Society in 1946, Arthur Holly Compton remarked that the bomb project revealed the extent to which "colored and white, Christian and Jew" could work together for a common purpose. Would it were the case that this kind of cooperation and collaboration could have continued after the war effort.

After World War II, a few white universities did begin to open up opportunities for blacks on their faculty as well as for blacks seeking graduate training within the departments. J. Ernest Wilkins was one of the first African Americans to hold a faculty appointment at the University of Chicago, for example. Still, the major problems facing blacks pursuing careers in science lingered—lack of access to a high-quality elementary and high-school science preparation, weak undergraduate curricula in certain black colleges, exclusion from admission to many white colleges, the high cost of graduate training, and systemic discrimination in the professional world of science. As an example of the last point, professional meetings of national scientific groups such as the AAAS were still being held in segregated cities like Atlanta and New Orleans, where, as late as the 1950s, black scientists who wanted to attend were not given living accommodations at the conference hotels. Several faculty at Fisk University signed a letter to *Science* magazine in 1951, protesting the action of the Mathematical Association of America in its denial of banquet tickets to black participants at a conference held at Vanderbilt in Nashville.

After World War II, the bulk of science education for blacks was carried out by African-American scientists who directed young talented blacks into the field while meeting the demands of their own research programs. Particularly noteworthy in this regard is the work of a chemist, Henry C. McBay (1914–96), who after doing extraordinarily well in his Ph.D. program at the University of Chicago, began in 1945 to teach at Morehouse College, the historically black college in Atlanta. Over the next thirty years his persistent guidance-principally of African-American men, but of women too-stimulated over forty blacks to obtain their Ph.D.'s in chemistry and allied fields. This represents one of the most stimulating stories of academic mentorship in the history of American science.

Issues of racial perspective aside, who goes into science, and who does not, has been in the past intertwined with the question of funding. In a direct way, blacks entering science in the first quarter of the twentieth century had no systematic form of support. The first important help with science per se came through fellowships of the National Research Council in the 1920s. In the case of blacks, these fellowships were supported by the Julius Rosenwald Fund, a private grant-awarding agency named for and established with an endowment donated by a wealthy Jewish-American businessman.

As early as the mid-1920s, the chairman of the National Research Council's Division on Biology and Agriculture talked about establishing endowments for science at black institutions to replace the somewhat piecemeal fellowship approach. The only other significant source of funding at the time came from the Rockefeller Foundation, which supported the two black medical schools (Howard University College of Medicine and Meharry Medical College) but did not set up major endowments there. Subsequently, the Macy Foundation, the Commonwealth Fund, since the late 1960s the Ford Foundation, and since the early 1970s the National Science Foundation (NSF) established special programs and thus filled significant voids in the support of African Americans pursuing scientific research. During the last two decades of the last century and continuing into the present, special efforts to distribute funds equitably have been made at both the NSF and the National Institutes of Health (NIH), but with only moderate success. African-American scientists still receive proportionately less than their representative share of government and private funding. The definition of "minority," once meant to include only blacks, has been broadened to encompass other underrepresented groups such as Hispanics and American Indians, while the budget for special programs for African Americans has not grown.

Much still remains to be understood about the intricacies of funding and its impact on African-American recruitment. To be sure, the system of peer review—with its emphasis on contacts, who you know, who knows you, etc.—plays an important role. The complicated scenarios underlying resource allocation require research that goes beyond the quantitative to a systematic interpretation of qualitative elements in grant proposals, peer reviews, and other documentation.. Life-history interviews with black scientists have often proved useful in identifying key issues regarding competitive funding practices, whereas, archival sources are crucial, if not unique in their role in constructing the elusive careers of earlier scientists like Bouchet.

After the passage of the 1964 U.S. Civil Rights Bill, new opportunities opened up somewhat for African Americans at both the undergraduate and graduate levels at many white American colleges and universities. Careers in the field of science became a firmer reality for black students, in both academia and industry. The 1970s and 1980s saw efforts by scientific organizations, universities, and learned societies to be more inclusive in their membership. The AAAS set up a program, "Opportunities in Science," to address the precise question of the underrepresentation of minorities and the handicapped in science and to study the issue of minority recruitment. The 1980s saw the leadership reins of AAAS taken up, for the first time, by an African American, Walter Massey (1938–   ), who became AAAS president in 1988.

In his subsequent position as director of the NSF (appointed 1990), Massey did not, however, speak out forcefully on the issue of the underrepresentation of blacks in scientific work or in the ranks of scientific institutions. Among the numerous examples one could cite is the near total absence of blacks from the elected membership of the National Academy of Sciences (NAS), the country's preeminent scientific organization. At present, there are only two or three elected Academy members who are black, out of a much larger number of qualified prospects.

The representation of African Americans in scientific careers continues to hover around 2–3% while blacks constitute around 12% of the total population. Increasing the pool will require time and effort of all concerned and represents a challenge for the new century. So far, recent intervention efforts have had a greater systematic impact on bringing women (another underrepresented group) into the ranks than on bringing blacks in, especially taking into account what has occurred with slots available in American colleges and universities. The promotion and academic tenure processes for blacks at white American universities point to difficulties similar to those found in the funding agencies. A major difficulty is peer review, which remains clouded in secrecy imposed by white faculty and administrators. Access to promotion and tenure dossiers is crucial for the social study of scientific careers of African American scientists.

Presently, conditions for African Americans in science leave much to be desired. While the racial integration of elementary and high schools since the 1950s and 1960s has supposedly opened up routes of access, a disguised form of segregation known as "tracking" has emerged—a system which assigns students to courses within a school system by ability groups. African-American students are often steered away by counselors and teachers, not all white, from the rigorous scientific and mathematical courses requisite for future training in science. When these students do survive high school and find themselves at prestigious white institutions, many are confronted with professors who have lower expectations for their performance than for the performance of white students.

Science as a discipline is cumulative, and training in science proceeds along well-established lines and routes, as much today as it did in Bouchet's day. It is not easy to overcome a sidestep along the way. Counseling and mentorship are therefore crucial in steering and encouraging black students to fulfill established goals in science in the same way that the early black scientists found and created their own networks of support. What is needed to increase the pool of African-American scientists is an aggressive effort of support, encouragement, and opportunity pursued in a systematic, comprehensive way all along the educational pipeline, from pre-school through graduate school. Only when full representation of blacks in science is achieved will the legacy of Bouchet's unique accomplishment as the first African American doctorate be realized. His was the start of a continuing goal in science for African Americans.

APPENDICES

Letter From a Former Student of Bouchet

1720 Taylor Street, N.W.
Washington, D. C. 20011
February 1, 1977

Dr. Ronald E. Mickens
Associate Professor of Physics
Fisk University
Nashville, Tenn. 37203

Dear Dr. Mickens:

In compliance with your request for any information I might have on Dr. Edward Alexander Bouchet, the first Black to receive a Ph.D. in the United States, I am herewith submitting as best I can my recollections of him. Although I was very young when I knew Dr. Bouchet, I feel honored to be one of the few persons living who remembers him.

I do not know the exact year that Dr. Bouchet came to Gallipolis, Ohio to be the Principal of the Lincoln School. My recollection is that he was well up in years when he came and that he remained for at least six or seven years until his health failed to the extent that he was forced to leave.

I recall hearing my parents and other members of the black community discuss some of the outstanding characteristics of Dr. Bouchet as follows- that he was a fine Christian gentleman, a consummate scholar, one who seemed very knowledgeable in all areas and yet was extremely modest and a person who set a wonderful example of politeness and graciousness for the community.

Members of the black community used to recount with pride Dr. Bouchet's first encounter with the Board of Education for the principalship of the Lincoln School (an all Black school). Even though the Board members were faced with the impressive qualifications of Dr. Bouchet such as: a Yale graduate, the Ph.D. degree (first Black in U.S. to receive one), the Phi Beta Kappa Honorary Membership (the first Black in the U.S. to achieve this honor) and various types of teaching experiences, they were still impelled to submit him to a "tour de force" type of written examination over the wide range of education which was expected to take a number of hours to complete. But Dr. Bouchet, the learned man that he was, finished very successfully this examination in a short time and spent the rest of the time looking out of the window. The result of this encounter was that Dr. Bouchet was awarded a life certificate for principalship in the state of Ohio. He was reputed by many to be the most brilliant educator in Southern Ohio.

At the time Dr. Bouchet was in Gallipolis I was in the elementary grades. When I reached the seventh and eighth grades Dr. Bouchet selected me as the student pianist for the School. I played for the High School music classes and for their Commencement Programs. Dr. Bouchet was often in charge of the music for these classes and it was here that I learned how extensive and vital his musical knowledge and leadership were.

Dr. Bouchet was a frequent visitor in our home. If my parents were living they could give a much better idea of his educational astuteness than I, because of my youth at that time.

I believe that Dr. Bouchet exerted a beneficial influence upon the achievements of my brother, J. Arnett Mitchell, a freshman in college when Dr. Bouchet came to our town. They had long conversations in the summers. My brother graduated with honors and the Phi Beta Kappa award from Bowdoin College, Burnswick, Me.

I am also of the opinion that Dr. Bouchet's selection of me to play for the high school music classes while I was still an elementary school pupil was perhaps one of the contributing factors which caused me to continue my education and achieve the Ph.D. in Higher Education.

Certainly it is impossible to assess the far reaching influence of Dr. Bouchet upon the hundreds of persons whose lives he touched.

I trust these recollections will be helpful to you.

I would appreciate it if you could tell who referred me to you for information.

Sincerely yours,

\ssn\

Lillian Mitchall Allen, Ph.D.

Retired Professor and Head of Music Education

College of Fine Arts, Howard University

Washington, D.C.

Appendix B

Willie Hobbs Moore
First African American Woman
Doctorate in Physics

Ronald E. Mickens*

Willie Hobbs was born 23 May 1934 in Atlantic City, New Jersey, the daughter of Bessie and William Hobbs. She obtained three degrees from the University of Michigan: the BSEE in 1958, the MSEE in 1961, and the Ph.D. in Physics in 1972. Hobbs' doctorate research was done under the direction of the noted infrared spectroscopist Professor Samuel Krimm and was focused on a theoretical vibrational analysis of secondary chlorides for polyvinyl-chloride polymers.

While at the University of Michigan, she also worked as a senior systems analyst at Datamax Corporation. Her career included engineering positions at Bendix Aerospace Systems Division, in Ann Arbor; Barnes Engineering Company, in Stamford, Conn.; and Sensor Dynamics Inc., in Ann Arbor, where she was responsible for the theoretical analysis of stress optical delays. As a research associate at the University of Michigan, she was responsible for establishing empirical models of optical hypersonic wakes which were used as diagnostic tools for verifying then existing flow field results. At KMS Industries she provided analytic support to the optics design staff. She also headed the Datamax analytic group which evaluated the performance of all the company's products. Finally, as an executive with Ford Motor Co., Moore worked with the warranty department of body and automobile assembly, in the reliability department, as a

*Clark Atlanta University, Department of Physics, Atlanta, GA 30314 (rohrs@math.gatech.edu).

manager of corporate learning. Her former colleagues credit her with having a significant role in spreading the use of Japanese-style engineering, and manufacturing methods at Ford in the 1980's. This was done in part by a widely circulated paper she wrote showing how the "robust" designs advocated by Genichi Taguchi, an award-winning Japanese consultant and engineering professor, could be achieved under realistic work conditions.

Dr. Moore published her research results in a wide variety of scientific journals including Journal of Molecular Spectroscopy, Journal of Chemical Physics, Journal of Applied Physics, Biopolymers, and Die Makromoleculare Chemie. In particular, she authored a series of papers with Krimm on the vibrational analysis of peptides, polypeptides, and proteins.

Moore was very active in a number of community clubs and organizations, particularly those concerned with the education of black youth. In particular, she contributed many of her efforts to tutoring and teaching at the Saturday African-American Academy in Ann Arbor. This was a community-run program which provided science and mathematics instruction for grades 5–12. Other memberships included Delta Sigma Theta, The LINKS, and Jack and Jill.

Dr. Willie Hobbs Moore had two sisters: Alice Doolin, a retired school teacher now living in Indianapolis, Indiana; and Thelma Gordy, a retired laboratory technologist residing in Atlantic City, New Jersey. Her husband Sidney L. Moore received a BS degree in mathematics and science eduction from Jackson State University (Mississippi) and an MS degree in educational psychology from Michigan University. For forty years (until 1997), he taught adolescents with psychological problems at the University of Michigan Hospital. Willie and Sidney had two children: a daughter, Dorian who received both undergraduate and medical degrees from the University of Michigan and, a son, Christopher who obtained the BS in physiology at the University.

After a fight with cancer, Dr. Moore died on Monday 14 March 1994 in her Ann Arbor home. At the 1995 National Conference of Black Physics Students she was awarded the Edward A. Bouchet Award posthumously.

Selected Research Publications

Moore, W. H. and Krimm, S., "Calculations of the Vibrational Spectrum of Crystalline Syndiotactic Poly(vinyl chloride)." *Bull. Amer. Phys. Soc.* **18**, 403 (1973).

Krimm, S. and Moore, W. H., "Defect Induced Infrared Absorptions in PVC." *Bull. Amer. Phys. Soc.* **18**, 403 (1973).

Moore, W. H., Ching, J. H. C., Warrier, A. V. R., Krimm, S. "Assignment of Torsion and Low Frequency Bending Vibrations of Secondary Chlorides." *Spectro. Chem. Acta.* **29A**, 1874–1858 (1973).

Moore, W. H., and Krimm, S., "A Complete General Valence Force Field for Secondary Chlorides." *Spectro. Chem. Acta.* **29A**, 2025–2042 (1973).

Moore, W. H. and Krimm, S., "Vibrational Analyses of 2,4-dichloropentane and 2,4,6-trichloroheptane." *J. Molec. Spectros.* **51**, 1–26 (1974).

Moore, W. H. and Krimm, S., "Intermolecular Interactions in Secondary Chlorides." *J. Chem. Phys.* **59**, 5195–5219 (1973).

Moore, W. H. and Krimm, S., "The Vibrational Spectrum of Crystalline Syndiotactic Poly(vinyl chloride)." *Die Makromoleculare Chemie*, Suppl. 491–506 (June 1975).

Hsu, S. L., Moore, W. H. and Krimm, S., "A Vibrational Analysis on Crystalline *Trans*-1,4-Polybutadiene," *J. Appl. Phys.* **46**, 4185–4193 (Oct. 1975).

Moore, W. H. and Krimm, S., "Transition Dipole Coupling in Amide I Modes of β-Polypeptides," *Proc. Nat. Acad. Sci. USA* **72**, 4933–4935 (Dec. 1975).

Hsu, S. L., Moore, W. H. and Krimm, S., "Vibrational Spectrum of the Unordered Polypeptide Chain: A Raman Study of Feather Keratin," *Biopolymers* **15**, 1513–1528 (Aug. 1976).

Moore, W. H. and Krimm, S., "Vibrational Analysis of Peptides, Polypeptides, and Proteins. I. Polyglycine I," *Biopolymers* **15**, 2439–2464 (Dec. 1976).

Moore, W. H. and Krimm, S., "Vibrational Analysis of Peptides, Polypeptides, and Proteins. II. β-Poly(L-alanine) and β-Poly(L-alanylglycine)," *Biopolymers* **15**, 2465–2483 (Dec. 1976).

Rabolt, J. F., Moore, W. H. and Krimm, S., "Vibrational Analysis of
    Peptides, Polypeptides, and Proteins. 3. α-Poly(L-alanine),"
    Macromolecules **10**, 1065–1074 (Sept.–Oct. 1977).

Elmer Samuel Imes*
Scientist, Inventor, Teacher, Scholar

Ronald E. Mickens†

Elmer Samuel Imes was the first black scientist to make a significant contribution to physics. His work had a major impact on the understanding and interpretation of quantum phenomena during the period from 1919 to 1925. He also made contributions to physics instrumentation through his construction and improvements to infrared spectrometers. During his lifetime, his research was extensively quoted and referenced in leading scientific journals in the United States and Europe by physicists and chemists studying the properties and molecular spectra of diatomic molecules.

The Imes family had its American roots in the southern central region of Pennsylvania. William L. Imes wrote in *The Black Pastures* that the family were "rugged farming folk" and had "free black ancestry running back several generations," even in the latter part of the nineteenth century. Elmer Samuel Imes' parents, Benjamin Albert Imes and Elizabeth Wallace, met in Oberlin, Ohio, and were married there. Benjamin graduated from Oberlin College in 1877 and in 1880 obtained his divinity degree from Oberlin Seminary.

Elmer Samuel Imes was born October 12, 1883, in Memphis, Tennessee. Two other brothers, Albert Lovejoy and William Lloyd, soon

---

*This article is from *Notable Black American Men*, Jessie C. Smith, editor (Gale; Farmington Hills, MI; 1999), pp. 589–591. Reprinted by permission of the Gale Group.

†Clark Atlanta University, Department of Physics, Atlanta, GA 30314 (rohrs@math.gatech.edu).

followed. William Lloyd became a prominent theologian and had a distinguished career serving as pastor in northern churches as well as dean of the chapel at Fisk University.

Imes attended grammar school in Oberlin, Ohio, from about 1889 1895 and the Agricultural and Mechanical College High School in Normal, Alabama, from about 1895 to 1899. He then enrolled at Fisk University, where he received the B.A. degree in 1903. For the next several years he taught physics and mathematics at Albany Normal Institute located in Albany, Georgia. He then returned to Fisk University around 1910, where he remained until 1915. During this time Imes completed academic work for the master of arts degree and also served as an instructor in science and mathematics.

With his M.A. degree, Imes reached the limits of what Fisk University could offer in terms of research. Since he was black, clearly any additional studies would have to be done at an institution outside the South. In 1915, he enrolled in the doctoral program in physics at the University of Michigan. Imes's academic work during his first year was of such high quality that he was offered a graduate fellowship for the remainder of his study in there.

At the University of Michigan, Imes began his research under the guidance of Professor Harrison M. Randall. Just prior to Imes's arrival, Randall had gone to Germany to work in Professor Friedrich Paschen's spectroscopy laboratory. There he concentrated on the production, characterization, and measurement of the infrared region of the spectrum. Returning to Ann Arbor, Randall and his students began to design and build infrared spectrometers of higher resolving power and to build more sensitive detectors. The most notable of his students was Imes.

New Field of Scholarship Opened

Gary Krenz in *LSA Magazine* cited Randall and Imes for publishing in 1919 a single work that ushered in a new field of research, "the study of molecular structure through the use of high-resolution infrared spectroscopy. Their work revealed for the first time the detailed spectra of simple-molecule gases, leading to important verification of the emerging quantum theory and providing, for the first time, an accurate measurement of the distances between atoms in a molecule."

Imes received the doctorate degree in physics in 1918, and his dissertation was published in the *Astrophysical Journal* in 1919. Some extensions of this work appeared in a short joint paper that Imes published with Randall in 1920 in *Physical Review*. The fundamental significance of Imes's research was clearly stated by Professor Earle Plyler in 1974:

> Up until the work of Imes, there was doubt about the universal applicability of the quantum theory to radiation in all parts of the electromagnetic spectrum. Some held that it was useful only for atomic spectra (electronic spectra); some held that it was applicable for all electromagnetic radiation. ... Imes's high resolution work on HCl, HBr and HF was the first clear-cut experimental verification of the latter hypothesis, namely, that the rotational energy levels of molecules are quantized as well as the vibrational and electronic levels. ... Thus, Imes's work formed a turning point in the scientific thinking, making it clear that quantum theory was not just a novelty, useful in limited fields of physics, but, of widespread and general application.

The significance of Imes's results was immediately recognized by major quantum scientists in both America and Europe. In the decade after 1919, his *Astrophysical Journal* paper would be extensively cited in research papers and reviews of the research literature on rotational-vibrational spectra of diatomic molecules. Within a very short time, discussions of Imes's work began to be incorporated into standard textbooks on modern physics. In each of these books, Imes's rotational-vibration spectrum of the fundamental absorption bands of HCl was prominently displayed. Imes's work also provided early evidence using molecular spectra for isotopes.

An interesting insight into how Imes was viewed by the scientific community can be obtained from a letter of professor Youra Qualls of Tuskegee Institute:

> I worked for Dr. Imes as secretary in either my junior or my senior year in college. At the time Dr. Imes was writing a history of physics. One of the delightful tasks I assumed was going through foreign science journals to note references to

the work of "Imes of the U.S.A." My French and German
were elementary by his standards, but I was able to keep up
with this chore reasonably well. I mention it only to say that
Dr. Imes was, I believe, far better known abroad than he was
in his own country.

To pursue the last sentence I will recall an incident occurring
several years after my graduation. I was teaching at Langston
University in the year that Dr. Charles S. Johnson became
Fisk's president. On my way to Fisk for the inauguration, I
met a Dr. Nielson, Dean of the Graduate School of Sciences
at the University of Oklahoma. As Dr. Nielson and I talked,
Dr. Imes' name came into the conversation. He told me that
he had become familiar with the work of "Imes of the U.S.A."
during his student days in Denmark but that he had never
known that Imes was a Negro.

In further recognition of his outstanding research achievements,
Imes was elected at the University of Michigan to membership in
Sigma Xi, a national honor society scientific research.

An Engineer and Intellect in New York

A year after receiving the doctorate, Imes married Nella Larsen on
May 3, 1919 in New York. Their marriage produced no children,
and they eventually divorced in Nashville, Tennessee, on August
30, 1933. Larsen was a gifted writer of the Harlem Renaissance.
In 1930 she was the first African American female creative writer to
win a Guggenheim Fellowship.

Imes spent the 1920s working in and around New York in several capacities as an engineer and applied physicist: as a consulting
physicist (1918–22); a research physicist with the Federal Engineers
Development Corporation (1922–24); with the Burrows Magnetic
Equipment Corporation (1924–27); and as a research engineer at
E. A. Everett Signal Supplies (1927–30). His applied research and
engineering activities resulted in four patents concerned with deter-
mining the electrical and magnetic characteristics of certain materials
and constructing instruments to measure these properties accurately.

During this New York period, Imes's own scholarly and literary interests, as well as his marriage to a well-known writer of the

Harlem Renaissance, brought him in contact with many members of the African American intellectual and elite throughout the 1920s. These figures included W. E. B. Du Bois, Charles S. Johnson, Arna Bontemps, Langston Hughes, Richard Nugent, Aaron Douglas, Walter White, and Harlem Renaissance benefactor Carl Van Vechten. Many of these same people would appear again in Imes's life during the 1930s through their connection to Fisk University.

The Return to Fisk University

The prevailing economic situation and possibly a conflict with his employer at the E. A. Everett Signal Supplies Company led Imes to consider returning to Fisk University at the close of the 1920s. After protracted negotiations with Thomas Elsa Jones, president of Fisk University, Imes returned to the university as professor and chair of the department of physics. He immediately began to reorganize the undergraduate physics curriculum and made preliminary studies for the initiation of a full-fledged graduate program.

During Imes's tenure at Fisk (1930–41), he and his students were involved in several research projects using X-rays and magnetic procedures to characterize the properties of various materials. Several students were sent to the University of Michigan during the summers to learn X-ray techniques and Imes spent at least one summer at New York University carrying out experiments on magnetic materials. He also continued his research in infrared spectroscopy and returned to the University of Michigan for several summers to conduct experiments on the fine structure of the infrared rotational spectrum of acetylene.

Imes was active in several professional societies including the American Physical Society, American Society for Testing Materials, and the American Institute of Electrical Engineers. He would often attend national meetings of these organizations in the larger northern cities or in Canada. Another measure of his standing in the scientific community is indicated by his listing in the sixth and later editions of *American Men of Science*.

A major concern of Imes's was the training of students, and a number of his students enrolled at the University of Michigan. James

R. Lawson earned a Ph.D. in 1939 for work focusing on infrared spectroscopy, and Lewis Clark graduated with an M.S. degree in physics.

As part of the Negro intellectual elite, Imes felt that the students at Fisk, as well as his friends and colleagues, should be exposed to the general outlines and themes of science. To this end, he developed a course called "Cultural Physics" and wrote a book-length manuscript to be used in it. The manuscript presents a general summary of the history of science, beginning with the Greeks and continuing up to the early part of the twentieth century.

In addition to his duties as chair of the department of physics, which included detailed work on the design of a new science building, Imes carried out an extensive correspondence with other researchers, equipment designers and manufacturers. He was also heavily involved with both the academic and social affairs of Fisk University, including being in charge of and running various films for the university community, participating in the planning and execution of the Annual Spring Arts Festival, and serving on various scholarship and disciplinary committees.

Imes's tenure at Fisk was marked by a national scandal over his involvement with a white administrator at Fisk. This situation, as well as other complications, eventually led to his divorce from Nella Larsen on August 30, 1933.

Near the end of the 1930s, Imes's health deteriorated, and he returned to New York to be treated for cancer. He died on September 11, 1941. Imes's obituary in *Science* was written by his friend and scientific colleague W. F. G. Swann:

> In the death of ... Imes science loses a valuable physicist, an inspiring personality and a man cultured in many fields. ... His thesis ... [dealt with] infra-red spectra, a subject on which he has acquired an international reputation ... It was the writer's privilege to become acquainted with Professor Imes in his graduate student days at the University of Michigan, where his research laboratory was a mecca for those who sought an atmosphere of calm and contentment. Imes could also be relied upon to bring to any discussion an atmosphere of philosophic soundness and level headed practicalness. ... In his passing, his many friends mourn the loss of a distinguished scholar and a fine gentleman.

References

Davis, Thadious M. *Nella Larsen: Novelist of the Harlem Renaissance.* Baton Rouge. Louisiana State University, 1994.

"Fisk Professor is Divorced by N.Y. Novelist," *Baltimore Afro-American,* October 7, 1933.

Fuson, Nelson. Notes taken of Dr. Earle Plyler's [Professor Emeritus of Physics, Florida State University] Symposium Talk at the Fisk Infrared Institute's 25th Anniversary Celebration, August 16, 1974.

Imes, Elmer S. "Measurements of the Near Infra-Red Absorption of Some Diatomic Gases," *Astrophysical Journal* 50 (1919): 251–76.

Imes, William Lloyd. *The Black Pastures.* Nashville: Hemphill Press, 1957.

Krenz, Gary D. "Physics at Michigan: From Classical Physics to Nuclear Research, 1888–1938," *LSA Magazine* 12 (Fall 1988): 10–16.

Qualls, Youra. Letter to Ronald E. Mickens, April 20, 1982.

Randall, Harrison McAllister and Elmer S. Imes. "The Fine-Structure of the Near Infra-Red Absorption Bands of the Gasses HCl, HBr, and HF," *Physical Review* 15 (1920): 152–55.

Swann, W. F. G. "Elmer Samuel Imes," *Science* 94 (26 December 1941): 600–601.

Collections

The papers of Elmer Imes are in the Carl Van Vechten Collection, Manuscript Division, New York Public Library; the Carl Van Vechten Collection, Beinecke Library, Yale University; and in the Special Collections Department, Fisk University Library. A manuscript copy of Imes's article "Measurements of the Near Infra-Red Absorption of Some Diatomic Gases" is also in the Fisk Collection.

The Genesis of the National Society of Black Physicists*

Ronald E. Mickens[†]

The following gives "my story" of the formation of what is now called the National Society of Black Physicists (NSBP). This chronology is based on my personal records/documents[1] and the first two newsletters of the Society of Black Physicists.[2,3]

As a graduate student in physics at Vanderbilt University, I attended my first Southeastern Section Meeting of the American Physical Society in 1966 at Clemson University. There, I met Howard Foster, the Chair of the Physics Department at Alabama A and M University. We formed an immediate friendship which lasted up to his premature death in the early 1970's. For several years, Howard collected the names and other significant information on Blacks having degrees in physics: Roster of Blacks in Physics. After his death, I continued this activity for another decade.

I completed my doctorate in 1968 and with a National Science Foundation Postdoctoral Fellowship spent the next two years at the Center for Theoretical Physics, MIT. One of the most interesting persons at the Center was James Young, then on leave from Los Alamos National Laboratory, who would soon become Professor of

*This article first appeared in the *Spring* 1999 *Newsletter* of the National Society of Black Physicists.

†Clark Atlanta University, Department of Physics, Atlanta, GA 30314 (rohrs@math.gatech.edu).

Physics at MIT. In addition to our friendship and mutual respect for each other's scientific accomplishments, our discussions would often turn to the senior physicists in the black college community who mentored several generations of students who then went on to achieve doctorates in physics. These "elders" served as role models, provided the required intellectual tools for success in graduate school, and gave (when needed) both emotional and financial support to their students. In early 1972, we decided to organize a gathering to honor three persons: Halson Eagleson (Howard University), Donald Edwards (North Carolina A and T College), and John Hunter (Virginia State College). All these individuals were well known in our community, were considered excellent teachers, and had trained large numbers of students who completed the requirements for advanced degrees in physics. Fisk University was selected as the site for this event. There were three reasons for this decision: first, by this time, I was a member of the Fisk physics faculty; second, Fisk had a long tradition in both physics education and research; and third, Nashville was a convenient location for travelers coming from both coasts of the country.

Jim Young and I asked Joseph Johnson III (Southern University) and Harry Morrison (on leave at Howard University from the University of California-Berkeley) to serve with us as an Awards Committee. However, it was understood that the detailed planning and related activities were to be done by me. A major time-saver in this effort was Howard Foster's Roster of Blacks in Physics. I wrote a letter explaining the purpose of the gathering in Nashville and requested a contribution of $50.00 per person to cover expenses of the affair.

At 5:30 p.m., 9 December 1972, approximately sixty friends, colleagues, and former students of the three guests of honor met at the Fisk University Faculty Club House for a pre-dinner social hour. The three awardees were interviewed in a separate room by representatives of the local print and broadcast press. Excerpts of these discussions, along with comments from others in attendance, appeared that night on two local television stations; the next day each of the newspapers published short articles on the event.

The Master-of-Ceremonies for The Awards Dinner for the First National Physics Fellows was Rutherford Adkins (Fisk University).

After an excellent meal, historical perspectives were given by James Lawson (Fisk University), Warren Henry (Howard University), and Harry Morrison. This was followed by the presentation of each recipient's biographical sketch by a former student and individual remarks by each awardee. The three each received a plaque and a certified check for $250.00. The citation text read as follows:

> In recognition of distinguished service to physics and society, we the undersigned present to _____ the first National Physics Fellows Award. This Citation is gratefully awarded to the aforementioned by a group of his colleagues and friends who observe that the black experience in physics and science generally has been enriched by his gift and humanity. Presented by the Awards Committee, Nashville, Tennessee, 9 December 1972.

The Second National Physics Award Ceremony was held at Howard University on May 1, 1975. The planning committee consisted of Anna Coble and Arthur Thorpe, both of the Howard University Physics Department, and myself. It was decided that the Awards Dinner would be preceded by a full day of formal scientific lectures. The scientific program is listed below:

Speaker	Title of Talk
*Walter Massey	"The Surface of Quantum Liquids"
*William Jackson	"Laser Induced Photo-Luminescence Spectroscopy"
*William Lester	"Theoretical Studies of Low Energy Inelastic Molecular Scattering"
*Ernest Coleman	"Research Advances in High Energy Physics"
*James Young	"Interactions: Prognosis for the Future"
*Warren Henry	"Historical Perspective on Magnetism"

The three awardees were Herman Branson, Warren Henry, and James Lawson. The Awards Dinner ceremony followed closely the format established at Fisk University. In particular, each awardee was presented with a citation plague and a certified check for $250.00. Several hundred persons attended the Awards Dinner.

The enthusiasm generated by the Fisk and Howard events led to a Day of Scientific Lectures and Seminars that was held the following year (April 1, 1976) at Morehouse College. The prime organizers were Carl Spight and myself. At the end of the meeting, representatives from Morgan State University volunteered to put on a similar program in 1977. Another important feature of the Morehouse meeting was that many discussions took place on the possibility of establishing some type of national black physics organization. The following are some of the persons who made significant contributions, in the period 1976–77, to the plans for creating the proposed organization: James Davenport, Warren Henry, Walter Massey, Harry Morrison, Carl Spight, and James Young.

The statements now to be presented provide a concise summary of what took place at the Morgan State University meeting (April 28, 1977) and the following meeting, held again at Morehouse College (March 29–30, 1978). These statements are excerpted from reference[2]:

"... The Society was inaugurated on Thursday, April 28, 1977 at Morgan State University, Baltimore, Maryland with interim structures and officers. The general purpose of the Society is to promote the professional well-being of black physicists within the scientific community and within society at large, and to develop and support efforts to increase the opportunities for, and numbers of, Blacks in physics. The Society is not in conflict with either the goals or the mission of the A.P.S. or the A.A.P.T. or any other of the mainstream professional organizations and is not intended to supplant any of them. Rather, the Society expresses the need for an organization in which Blacks play a major role in creating and developing activities and programs themselves for themselves. ..."

"... The first business meeting of the Society of Black Physicists was held in the early afternoon of Friday, March 31, 1978 at Morehouse College, Atlanta, Georgia at the end of the Fifth Annual Day of Scientific Lectures. ... At the business meeting reports were given by Walter Massey and James Davenport who served ably on an interim basis as, respectively, Society president and secretary-treasurer. ..."

The Society then elected its first full-time officers:

President: Carl Spight (Morehouse College),
Treasurer: Walter Massey (Brown University),
Exec. Member: James Davenport (VA State College).

These three individuals constituted the executive committee and were[2] "charged with the following short-term activities:

a) Drafting of a formal statement of purpose and Bylaws of the Society ...
b) Continuing the membership drive ...
c) Continuing the compilation of the roster of black physicists (under the direct supervision of Ronald Mickens) ...
d) Initiating a Society "newsletter" ...
e) Establishing liaison with the Minorities Committee of the AAPT and APS ...
f) Representing the Society to all meetings of the (newly formed) Council of Black Scientific and Technical Organizations ...
g) Continuing the annual Day of Scientific Lectures and Banquet....."

Two important points should be noted. First, the original name of the organization was the Society of Black Physicists. Second, the 1978 meeting at Morehouse College was called the Fifth Annual Day of Scientific Lectures. There existed in the thoughts of many one major line of reasoning for using "fifth" in the title of this meeting. The new Society was about to involve itself with other national organizations on a variety of issues related to both minority science education and the full participation in the scientific affairs of this nation. It was felt that in the deliberations to come and in the search for funds to support Society projects, advantages would accrue from having an organization with a "history." Consequently, starting with the first ceremony at Fisk and counting the Howard ceremony as second, it follows that the meetings at Morehouse (1976), Morgan (1977), and Morehouse (1978), would, respectively, be the third, fourth, and fifth Day of Scientific Lectures! Following this "logic," the second meeting of the Society of Black Physicists was held at Knoxville College[3] along with the Sixth Annual Day of Scientific Lectures during 26–27 April 1979.

References

1. R. E. Mickens, personal records.
2. C. Spight (Morehouse College), editor, *Society of Black Physicists Newsletter*, Volume 1, Number 1, June 1978.
3. C. Spight (Morehouse College), editor, *Society of Black Physicists Newsletter*, Volume 1, Number 2, April 1979.

Appendix E

The Bouchet Institute

I. First Edward Bouchet International Conference on Physics and Technology*

The first Edward Bouchet International Conference on Physics and Technology was held in Trieste, Italy June 9–11, 1988. This conference was the joint effort of the International Centre for Theoretical Physics (ICTP) and Black American Friends of ICTP (BAF/ICTP), a support group created for this purpose. The motivation for the meeting was to provide Black physicists from America and African physicists from African countries an arena in which to: share their research results; discuss current topical issues in physics; address problems of mutual concern; and create a continuing organization. The conference consisted of scientific paper presentations, topical sessions on scientific issues of current interest, and organizational workshops, formal and informal, aimed at the evolution of a long term relationship. The conference was named in honor of Edward A. Bouchet who became, in 1876, the first Black American Ph.D. recipient and the first known person of African descent to earn the Ph.D. degree in physics. There were more than 50 participants.

The idea for such an interaction and the invitation to Trieste came from Prof. Abdus Salam, Nobel Laureate and ICTP Director.

---

*This "press release" (dated 26 January 1989) was prepared by Professor Joseph A. Johnson III, conference co-convener, and Dr. Lynette Edmonds Johnson, conference chairperson. At that time both were at The City College of New York in the Department of Physics.

BAF/ICTP was organized by physicists (NSBP). Profs. Bennie Ward (a participant at ICTP) of the University of Tennessee and Joseph Johnson III (a participant in the ICTP sponsored International Network for Applied Physics) of CCNY agreed to be the Black American co-conveners of this organization. In November, 1987, ICTP formally requested that BAF/ICTP convene the conference in keeping with the format established for the annual meetings of NSBP. The population targeted for initial participation included recent contributors to NSBP meetings and African physicists in residence and/or easily accessible to ICTP at the time designated for the conference. An advisory committee was formed as follows: Prof. J. A. Johnson III (The City College of New York), Prof. B. F. L. Ward (Univ. of Tenn.), Dr. L. E. Johnson (The City College of New York), Dr. A. M. Johnson (AT&T Bell Laboratories), Prof. S. C. McGuire (Alabama A&M University) and Dr. M. D. Slaughter (Los Alamos National Laboratory). Prof. Jean-Pierre Ezin of ICTP was designated as Conference Coordinator. Dr. Lynette E. Johnson of The City College was designated as Conference Program Chairperson.

The conference was supported financially by ICTP, the National Science Foundation (through a grant to Dr. L. E. Johnson) and the various home laboratories and institutions of the participants. It was hoped that this conference would provide a unique new vehicle for: the dissemination of research results; the development of new ideas, insights and activity at the current frontiers in technology; and the general scientific enrichment of the world-wide community of physicists of African descent.

From this conference, a new entity has indeed been created, specifically, the Bouchet Institute/ICTP. This will be an umbrella organization serving primarily as a conduit and implementer of six programs:

(1) Visiting Scholars and Collaborations (Af→USA);
(2) Equipment Transfer and Collaborations (USA→Af);
(3) Visiting Lecturers and Collaborations (USA→Af);
(4) Student Research Facilitation and Guidance (Af→USA);
(5) The 2nd Bouchet Conference (for 1990 in Africa);
(6) Fund-raising for some of the above (USA→(Af,USA)).

The actual activities of the Bouchet Institute will take place through program committees, created and structured by the Council of the Bouchet Institute, with membership including Council members and other interested persons (not necessarily on the Council) as appropriate. On September 30, 1988, Prof. Salam formally created the Bouchet Institute/ICTP with the following appointments:

(A) Members of the First Council of the Bouchet Institute/ICTP
Professor Francis Allotey (U. S. T. Kumasy, Ghana)
Dr. Charles Brown (AT&T Labs, Georgia)
Professor Gallieno Denardo (ICTP)
Professor James Ezeilo (University of Nigeria, Nsukka)
Dr. Jean-Pierre Ezin (University Nationale du Benin, Cotonou)
Professor Mohamed Hassan (TWAS, Trieste, Italy)
Dr. Anthony Johnson (AT&T Labs, New Jersey)
Professor Joseph Johnson (The City College, CUNY)
Professor Ronald Mickens (Atlanta University)
Dr. Leonard Shayo (University of Dar-es-Salaam, Tanzania)
Dr. Milton Slaughter (Los Alamos National Laboratory)
(B) Members of the Executive Committee
F. Allotey, G. Denardo, J. P. Ezin, A. Johnson, J. Johnson, L. Shayo, M. Slaughter.

The first meeting of the council and the Executive Committee will take place in April 1989 at the ICTP in Trieste, Italy.

II. First Bouchet Council Meeting

The first meeting of the Advisory Council of the Bouchet Institute was held in Trieste, 13–15 April 1989. This is a summary of the actions taken.

The Council decided that the name of the Institute shall be:
Edward Bouchet-ICTP Institute.
The programme of the Institute is as follows:

(i) Collaborations between Black American and African Physicists, including collaborations involving the transfer of equipment;

(ii) Visits by African students to American universities for the purpose of completing their graduate studies;

(iii) Visits by African scholars to American universities during the course of their research;

(iv) Biennial Bouchet Conferences on themes of mutual interest of importance to the development on the African continent.

The collaborations should, generally, extend over two or three years and involve regular visits by Black American physicists with their African colleagues. The African students, pursuing their research at American universities, would perform experimental or theoretical investigations in Physics, Technology and/or Mathematics under the joint supervision of both their African and Black American co-mentors; the final stages of the graduate work would be completed at their home (African) institution. The visits by Africans scholars would be facilitated by their Black American colleagues.

The Council decided to convene the 2nd Bouchet International Conference in Accra, Ghana from 14–17 August 1990. Details concerning the structure and schemes of this conference are indicated in the attached call for papers. A programme planning sub-committee, consisting of the members of the executive committee, will meet in Accra, from 12–13 March 1990, in order to make final plans for the conference. The Council decided that the City College, CUNY (USA) will serve as a funding implementer for support from U.S. government sources in connection with the Institute.

In general discussion, the Council agreed to include opportunities for the transfer of library materials and for equipment gifts through TWAS/ICTP in general informational mailings. The Council agreed to prepare a brochure on the Bouchet Institute, also to be included in general informational mailings. The Council explored various possible initiatives which could be undertaken with regard to the long term development of support resources, viz, funding. Broadly stated, after some deliberations, an overall long term annual operating budget for the Institute of US$ 1.5 million was estimated. Specific activities and responsibilities were assumed by each Council member for programme development, solicitation of funds (short term and long term), and for implementing the decisions of the Council. During the course of the Council Meeting, a representative group of African scientists at ICTP met the Advisory Council and shared concerns, criticisms and hopes for the future development of physics and technology on the African continent.

The Council decided that its next meeting will be on the afternoon of 17 August 1990, in executive session, at the site of the 2nd Bouchet Conference.

The Council determined the following structure for its continuing activities:

Professor J. A. Johnson III, Chairman

Dr. J. P. Ezin, Vice Chairman

Dr. L. K. Shayo, Vice Chairman

Professor G. Benardo, ICTP Liaison

Specific duties associated with these titles, except those immediately implied by the titles themselves, will be determined in due course as the need arises.

III. The Constitution for the Edward Bouchet-Abdus Salam Institute

The EBASI Constitution was adopted in Trieste, Italy on 9 November 2000.

Preamble

- Realizing the importance and role of science and technology, and in particular that of Physics and Mathematics, in the development of Africa;
- Considering the challenge, posed by the late Nobel Laureate Physicist, Professor Abdus Salam, to the population of African-American Physicists and Mathematicians in the USA, to contribute towards the promotion of Physics and Mathematics in Africa;
- Recognizing the fact that Edward A. Bouchet was the first person of African descent to receive the Ph.D. degree in Physics;
- Considering the successes already achieved by EBASI over the period of 10 years notwithstanding its small size as well as the financial limitations;
- Considering the untapped potential of the African-American

Physicists in the USA to realize the broad objectives of the EBASI in collaboration with African Physicists and Mathematicians in Africa; and

- Honoring both Edward A. Bouchet and Abdus Salam;

the EBASI Council declares this Constitution to guide the operation of the Institute from the date of its adoption and as amended from time to time by the EBASI Council.

Article 1

Name and Headquarters

The Edward Bouchet-Abdus Salam Institute, hereinafter referred to as the "Institute," is a non-profit making, international, non-governmental organization.

The headquarters of the Institute shall be located at the Abdus Salam International Center for Theoretical Physics, Trieste, Italy.

Article 2

Objectives

A. General Objectives

The general objective of the Institute shall be the promotion of science and technology, and in particular Physics and Mathematics, in Africa with a view to strengthen the role of African scientists in the development of Africa through directing their scientific activities to the improvement of the quality of life of the African people. The Institute shall achieve this by promoting research collaboration between African-American Physicists and mathematicians in the USA with their counterparts in Africa, and by organizing conferences, schools and workshops in Africa and the USA to address specific development issues of greater relevance to the African continent.

B. Specific Objectives

1. To run sandwich programs for the training of African post-graduate students particularly at the level of Ph.D. under the general principle of the home institution offering the degree;
2. To encourage staff exchange between USA and African Universities that have staff who are participants of EBASI;
3. To organize schools, conferences and workshops on selected topics that have relevance to the development of Africa;
4. To promote collaboration and communication among African and African-American Physicists and Mathematicians, and with the International 'scientific community as a whole;
5. To increase scientific productivity of African and African-American Physicists and Mathematicians;
6. To popularize and promote science and technology particularly for the development of Africa;
7. To encourage international institutions and the private sector to support EBASI in particular, and in general increase their support in the promotion of science and technology for the development of Africa;
8. To do any other thing that will promote development in Africa by using science and technology.

C. Activities

The objectives of the Institute shall be met through the following activities to be sponsored by the Institute:

1. The creation of a data bank on the academic potential of African Universities and American Universities that have African-American Physicists and Mathematicians;
2. The organization of sandwich training programs particularly leading to the award of Ph.D. degrees by African Universities;
3. The organization of schools, workshops, and conferences on topics of greater relevance to the development of Africa;
4. The Support of staff exchanges between Universities/Research institutions in Africa and the USA, involving African-American

Physicists and Mathematicians going to Africa and African Physicists and Mathematicians going to the USA;

5. The Acquisition of research equipment, books and Journals from Universities and Institutions in the USA and providing such equipment, books and Journals to Universities and Research institutions in Africa;

6. The forging of close contacts with Universities, institutions and organizations all over the world for the purpose of furthering the objectives of EBASI;

7. The use of the opportunities available at the Abdus Salam ICTP, the International Center for Science and Technology of UNIDO, and the Third World Academy of Sciences for the promotion of the objectives of EBASI;

8. The undertaking of fundraising missions particularly in the USA in order to promote the objectives of the Institution.

9. The establishment of regional Centers of excellence in Physics and Mathematics in Africa.

10. The initiation of any other activities that will further the objectives of the Institute.

Article 3

Membership

There shall be four types of members of the Institute:

A. Council Members

These are the individual members of the EBASI Council originally appointed by the late Professor Abdus Salam in 1988 and any other individual that will be appointed by the Council as a self-sustaining body from time to time.

B. Members

These are individual Physicists and Mathematicians working in Universities and Research Institutions and appointed by the EBASI

Council in recognition of their demonstrated commitment to further the objectives of EBASI on a continuous basis.

C. Associate Members

These are individuals appointed by the EBASI Council in recognition of their demonstrated commitment to further the objectives of EBASI on a continuous basis.

D. Affiliated Institutions

These are institutions designated by the EBASI Council in recognition of their demonstrated support to EBASI on a continuous basis.

Article 4

Structure, Governing Body and Duties

The decision-making body of the Institute shall be:

A. The Council

The EBASI Council shall be the highest decision-making body of the Institute. The Council shall lay down the guiding principles and policies of the Institute for the purpose of enabling the Institute to further its objectives in the most efficient and elegant manner. As appropriate the Council shall elect a Chairperson.

B. The Executive Board

This shall be appointed by the Council from time to time, as a continuing entity to direct, as may be appropriate, the affairs of the Institute between sessions of Council Meetings. The Executive Board shall provide a report of its activities to the next meeting of the EBASI Council. The Chair of the Council shall serve as Chairperson of the Executive Board.

Article 5

Finances of the Institute

The funds of the Institute shall be obtained from:

1. Subventions and donations;
2. Contributions to support specific activities.

Article 6

Amendments to the Constitution

This Constitution can only be modified by the EBASI Council. Any modification shall be made at any of the regular meetings of the Council and passed by a two-thirds majority.

Appendix F

Selected Bibliography of Materials On African American Scientists

Russell L. Adams, *Great Negroes: Past and Present* (Afro-Am Publishing Co., Inc.: Chicago, IL; 1984; 3rd edition).

Richard Bardolph, *The Negro Vanguard* (Negro Universities Press; Westport, CT; 1971).

Silvio A. Bedini, *The Life of Benjamin Banneker: The Definitive Biography of the First Black Man of Science* (Charles Scribner's Sons, New York, 1972).

Hattie Carwell, *Blacks in Science: Astrophysicist to Zoologist* (Exposition Press; Hicksville, NY; 1977).

Empak Enterprises, Inc. (520 N. Michigan Ave., Suite 1004; Chicago, IL; 60611.)

    a) Vol. I: *A Salute to Historic Black Women* (1984);
    b) Vol. II: *A Salute to Black Scientists and Inventors* (1985);
    c) Vol. III: *A Salute of Black Pioneers* (1986).

Famous Black People in American History, Edu-Cards No. 275, (KPB Industries; Bethlehem, PA 18071).

Harry W. Greene, *Holders of Doctorates Among American Negroes: An Education and Social Study of Negroes Who Have Earned Doctoral Degress in Course*, 1876–1943 (Crofton; Newton, MA; 1974).

Louis Harber, *Black Pioneers of Science and Invention* (Harcourt Brace Jovanovich, New York, 1970).

Robert C. Hayden, *Seven Black American Scientists* (Addison-Wesley; Reading, MA; 1970).

Robert C. Hayden, *Eight Black American Inventors*, (Addison-Wesley; Reading, MA; 1972).

Claudia Henrion, *Women in Mathematics* (Indiana University Press; Bloomington, IN; 1997).

Portia P. James, *The Real McCoy: African-American Invention and Innovation*, 1619–1930 (Smithsonian Institution Press; Washington, DC; 1989).

James M. Jay, *Negroes in Science: Natural Science Doctorates*, 1876–1969 (Balamp Publishing, Detroit, 1971).

Edward S. Jenkins, "Impact of Social Conditions: A Study of the Works of American Black Scientists and Inventors," *Journal of Black Studies*, Volume 14 (#3), pps. 477–491 (1984).

Edward S. Jenkins, *To Fathom More: African American Scientists and Inventors* (University Press of America, New York, 1996).

James H. Kessler et al. (editors), *Distinguished African American Scientists of the 20th Century* (Oryx Press; Phoenix, AZ; 1996).

William M. King, "The Afro-American Scientist and Inventor: A Resource Bibliography," *Journal of Social and Behavioral Sciences* Volume 33 (#3), pps. 177–192 (1987).

Aaron E. Klein, *The Hidden Contributors: Black Scientists and Inventors in America* (Doubleday, New York; 1971).

Kristine Krapp (editor), *Notable Black American Scientists* (Gale Group, Detroit, 1999).

Kenneth Manning, *Black Apollo of Science: The Life of Ernest Just* (Oxford University Press, New York, 1983).

Kenneth R. Manning, "The Complexion of Science," *Technology Review*, November/December 1991, pps. 61–69.

Charles Meyer et al., *On the History of Physics at Michigan* (Department of Physics, University of Michigan; Ann Arbor, MI; 1988).

Ronald E. Mickens (editor), *The African American Presence in Physics* (Atlanta, GA; March 1999).

Alfred A. Moss, Jr., *The American Negro Academy* (Louisiana State University Press; Baton Rouge, LA; 1981).

National Science Teachers Association, *American Black Scientists and Inventors* (Washington, DC; 1975).

Virginia K. Newell et al. (editors), *Black Mathematicians and Their Works* (Dorrance; Ardmore, PA; 1980).

Willie Pearson, Jr., *Black Scientists, White Society, and Colorless Science: A Study of Universalism in American Science* (Associated Faculty Press, New York, 1985).

Willie Pearson, Jr. and H. Kenneth Bechtel (editors), *Blacks, Science, and American Education* (Rutgers University Press; New Brunswick, NJ; 1989).

SAGE: A Scholarly Journal on Black Women, Special issue on "Science and Technology;" R. E. Mickens (editor); Vol. VI, #2, Fall 1989.

Nancy N. Soper et al., *Physics Doctorates of Yale* (Department of Physics, Yale University; New Haven, CT; 1976).

Julius H. Taylor, et al., (editors), *The Negro in Science* (Morgan State College Press, Baltimore, 1955).

Wini Mary Edwina Warren, *Hearts and Minds: Black Women Scientists in the United States*, 1900–1960. (UMI Microform 9805375, UMI Company; Ann Arbor, MI; 1997).

Raymond B. Webster (editor), *African American Firsts in Science and Technology* (Gale Group, Detroit, 1999).

Herman A. Young and Barbara H. Young, *Scientists in the Black Perspective* (The Lincoln Foundation; Louisville, KY; 1974).

Lisa Yount, *Black Scientists* (Facts on File, New York, 1991).

THE AUTHORS

H. Kenneth Bechtel

Dr. Bechtel was born and grew up in Fargo, North Dakota. He received his undergraduate degree in sociology from North Dakota State University (B.A., 1972) and continued his education with graduate training in sociology at North Dakota State University (M.A., 1974) and Southern Illinois University at Carbondale (Ph.D., 1983). In 1982, he joined the Wake Forest University Sociology faculty where he has conducted research and published in the area of policy history, professional deviance, and the historical presence of minorities in science. In 1989, he co-edited (with Willie Pearson, Jr.) and wrote the introductory chapter for *Blacks, Education, and American Science: Perspectives on the Black Presence in American Science*. In 1995, he published *State Police in the United States: A Socio-Historical Analysis*, a book describing and analyzing the rise of state-level law enforcement agencies in the United States. Most recently, he has written "Edward Alexander Bouchet" for the *American National Biography* (1999), and "Scientific Deviance" for the *Encyclopedia of Criminology and Deviant Behavior* (2000).

Kenneth R. Manning

Dr. Manning is the Thomas Meloy Professor of Rhetoric and of the History of Science at the Massachusetts Institute of Technology. He has three degrees from Harvard University (B.A., 1970; M. A., 1971;

Ph.D., 1974) and has been at MIT since 1974. For the past two decades he has been involved in a major research project on "Blacks In American Medicine, 1860–1980." This work was funded by the Alfred P. Sloan Foundation, the Josiah Macy, Jr. Foundation, and the Henry J. Kaiser Family Fund. His book, *Black Apollo of Science: The Life of Ernest Just* (Oxford University Press, 1983) won The Pfizer Book Award of the History of Science Society (1984) and the Lucy Hampton Bostick Book Award (1984). It was also a finalist for the National Book Critics Circle Award (1984), the Robert F. Kennedy Book Award (1984) and the Pulitzer Prize in biography (1984). Dr. Manning has been a Keynote speaker at numerous universities and public events throughout the United States and abroad on topics in the history of science, the history of medicine, African Americans in science and medicine, and science education. Currently, he is working on a book manuscript on health care for African Americans and the role and experience of blacks in the American medical profession, 1860–1980; and the setting up of a database of associated historical collections.

Ronald E. Mickens

Dr. Mickens is the Distinguished Fuller E. Callaway Professor of Physics at Clark Atlanta University. He received his undergraduate degree in physics from Fisk University (B.A., 1964) and a doctorate in theoretical physics from Vanderbilt University. He has held postdoctoral positions at the Center for Theoretical Physics–MIT, The Joint Institute for Laboratory Astrophysics at the University of Colorado–Boulder, and Vanderbilt University. He has published more than 220 research papers, written five books and edited five volumes in the areas of nonlinear oscillations, applied mathematics, numerical analysis, and the history/sociology of African Americans in science. He serves on the editorial boards of several research journals including the *Journal of Difference Equations and Applications*. His professional memberships include the American Association for the Advancement of Science, the American Mathematical Society, the American Physical Society (for which he is a Fellow), The National Society of Black Physicists, and the History of Science Society.

In 1999, he was project director of an exhibit, "The African American Presence in Physics," which was displayed at a number of sites in the United States. He also created a wall poster and book associated with the exhibit, each having the same title as the exhibit.

Curtis L. Patton

Dr. Patton received his undergraduate degree in zoology from Fisk University (1956) and two graduate degrees (M.S., 1961; Ph.D., 1966) in microbiology from the University of Michigan. His honors and awards include a United States Public Health Service postdoctoral fellowship (1967–70) at the Rockfeller University and a U.S.P.H.S. Career Development Award at Yale University (1972–77). He is a Fellow of the Society of Infectious Diseases and received an Honorary Doctor of Science Degree (1991) from Fisk University. In addition to serving on a large number of national and internal advisory committees, he is at the Yale University School of Medicine the Chair of the Committee on International Health and Director of International Medical Studies. Currently, Dr. Patton is a professor in the Department of Epidemiology and Public Health.

Thomas Rodd, Jr.

Thomas ("Tim") Rodd, Jr., began his career at Hopkins School in 1971 as an English teacher and coach, and served as the school's headmaster from 1989 to 1999. He was educated at Yale (B.A., 1965) and Columbia (M.A., 1971). During the academic year, 1978–79, he was a Klingenstein Fellow at Teachers College, Columbia University, studying the history of composition teaching. In addition to Hopkins, he has worked at four other independent schools, among them elementary, secondary, day, and boarding schools. He now resides in New Hampshire.

John A. Wilkinson

John A. Wilkinson has worked in university and independent school administration for over forty years. A graduate of Yale College and Yale Graduate School, he served his *alma mater* in the roles of Associate Dean of Yale College, Dean of Undergraduate Affairs, Vice President for Development, and Secretary of the University. He also has been Headmaster and teacher of history at the Hopkins School (New Haven, CT), Germantown Friends School (Philadelphia, PA), and Portsmouth Abbey School (Portsmouth, RI). He currently is Vice President for Development and External Affairs of the American University of Beirut.

Mr. Wilkinson is an active member of the Headmasters Association, the Country Day School Headmasters Association, Head Mistresses of the East, the Yale Club of New York City, Social Science Club of New Haven, and the New Haven Colony Historical Society. He is former Chairman of the New Haven Development Commission.

An amateur historian, his interest in Edward Bouchet began with the discovery of Bouchet's photograph in the yearbook of the Hopkins School for the Class of 1866. During his tenure as Secretary of Yale, he commissioned a portrait of Bouchet, which now hangs in the Sterling Memorial Library of Yale.